Springer Theses

Recognizing Outstanding Ph.D. Research

For further volumes:
http://www.springer.com/series/8790

Aims and Scope

The series "Springer Theses" brings together a selection of the very best Ph.D. theses from around the world and across the physical sciences. Nominated and endorsed by two recognized specialists, each published volume has been selected for its scientific excellence and the high impact of its contents for the pertinent field of research. For greater accessibility to non-specialists, the published versions include an extended introduction, as well as a foreword by the student's supervisor explaining the special relevance of the work for the field. As a whole, the series will provide a valuable resource both for newcomers to the research fields described, and for other scientists seeking detailed background information on special questions. Finally, it provides an accredited documentation of the valuable contributions made by today's younger generation of scientists.

Theses are accepted into the series by invited nomination only and must fulfill all of the following criteria

- They must be written in good English.
- The topic should fall within the confines of Chemistry, Physics, Earth Sciences, Engineering and related interdisciplinary fields such as Materials, Nanoscience, Chemical Engineering, Complex Systems, and Biophysics.
- The work reported in the thesis must represent a significant scientific advance.
- If the thesis includes previously published material, permission to reproduce this must be gained from the respective copyright holder.
- They must have been examined and passed during the 12 months prior to nomination.
- Each thesis should include a foreword by the supervisor outlining the significance of its content.
- The theses should have a clearly defined structure including an introduction accessible to scientists not expert in that particular field.

S. J. Ashton

Design, Construction and Research Application of a Differential Electrochemical Mass Spectrometer (DEMS)

Doctoral Thesis accepted by
the Technical University of Munich, Germany

Springer

Author
Dr. S. J. Ashton
Department of Chemistry
University of Copenhagen
Copenhagen
Denmark

Supervisor
Prof. Dr. Matthias Arenz
Department of Chemistry
University of Copenhagen
Copenhagen
Denmark

ISSN 2190-5053 ISSN 2190-5061 (electronic)
ISBN 978-3-642-30549-8 ISBN 978-3-642-30550-4 (eBook)
DOI 10.1007/978-3-642-30550-4
Springer Heidelberg New York Dordrecht London

Library of Congress Control Number: 2012941303

Printed on acid-free paper

Springer is part of Springer Science+Business Media (www.springer.com)

Parts of this thesis have been published in the following journal articles

Ashton, S.J., Arenz, M. (2012) "Comparative DEMS Study on the Electrochemical Oxidation of Carbon Blacks" Journal of Power Sources (accepted Jun. 2012)

Ashton, S.J., Arenz, M. (2011) "A DEMS study on the electrochemical oxidation of a high surface area carbon black" Electrochemistry Communications 13(12): 1473–1475

Electrochemistry is the science which deals with the conversion of matter to electricity; and/or electricity to matter.

Kyvstiakovsky (L. Antropov, Theoretical Electrochemistry) 1910

Supervisor's Foreword

Sean Ashton began his doctoral thesis in my Emmy Noether electrocatalysis research group in the Department of Physical Chemistry at the Technische Universität München in late 2007, where he was given the challenge of designing and constructing a so-called DEMS (Differential Electrochemical Mass Spectrometer) instrument and applying it to investigate proton exchange membrane fuel cell (PEMFC) catalysts. His work and corresponding doctoral thesis were exceptionally well formulated, and he has provided my laboratory with a beautiful designed, effective research tool.

In this thesis, Sean first presents the broader significance of the electrocatalysis research topic of my group (now part of the University of Copenhagen), which concerns itself with the fundamental investigation and development of fuel cell electrocatalyst materials. He then presents and discusses the prior DEMS instrument designs and their research applications, carefully compiling and evaluating the crucial information contained in prior DEMS literature. The thesis then documents, in detail, the design of the instrument he constructed (which does not simply replicate earlier instruments) presenting and discussing the design considerations of both hardware and software. After constructing a functional instrument, Sean did not impetuously dive into performing research studies but instead took the time to develop methods of characterising and optimising the various components, which are documented in his thesis. Finally, the research capability of the DEMS instrument is then demonstrated, revealing new insights in two key areas of fuel cell related research. The first study concerns methanol oxidation, whereby the stark contrast in the potential dependent conversion of methanol to CO_2 on Pt and PtRu catalysts is clearly revealed. The second study involves the investigation of industrial fuel cell catalyst support oxidation, where DEMS is able to evaluate the quantity and rate of the electrochemical oxidation of carbon supports, separating complete and partial oxidation currents.

Overall, the information that Sean presents in this thesis, which is largely unpublished, is not only exceptionally useful to those wishing to construct a similar instrument, but is also of great interest because of the insights and their context that were obtained.

Copenhagen, Denmark, March 2012 — Matthias Arenz

Acknowledgments

I would first and foremost like to thank Prof. Matthias Arenz, whom will likely recall that in the interview and presentation for the Ph.D. position, I had to rather awkwardly respond to almost all questions regarding my master thesis with "I'm really sorry but I can't say. It's confidential". Despite that, he decided to patiently guide me through my work and moulded me into an 'electrochemist' for which I am truly grateful.

This work, however, would not have even begun on the right track if it was not for the kind advice of Dr. Zenonas Jusys, whose knowledge and experience in DEMS is unrivalled. I would also like to thank Dr. Karl J.J. Mayrhofer whose enthusiasm provided me with a firm foundation in electrochemistry, and also Dr. Gustav Wiberg because when he wasn't occupied finishing his own work, or singing in a choir (or the lab...), he was most likely teaching me Labview programming; without his influence I would almost certainly still be treating and analysing data today. There are also the remaining members of the electrochemistry group whom supported my work, namely Katrin and Markus to thank, particularly for the good-times we had setting up the laboratory in Copenhagen with Neuhavn and The Moose being special highlights. I would also like to express my appreciation to Prof. Ulrich Heiz and his research group for the pleasant working environment, the TU-München Mechanische Werkstatt for their meticulous fabrication of the DEMS components, and the Elektronik Werkstatt for regularly letting me 'borrow' things.

Of course, the completion of this work was unachievable without the support of my friends and family. I owe an awful lot to Gustav, Martin and Aras for providing mutual distraction from the highs and lows of our Ph.Ds. It was always great to get away from the relentless grinding of vacuum pumps, whether it was off to a Biergarten, the calmness of a snooker table, or even a trip to the ski slopes. Finally, I would like to especially thank my Mum and Dad, Jason and Jasmine for their unconditional love and support, and Bärbel, Steffi, Marga and Horst for their unreserved kindness. Thank you all.

Acknowledgments

Thank you all.

Contents

Abbreviations

AO	Analogue output
AFM	Atomic force microscopy
ATR-FTIR	Attenuated total reflection fourier transform infrared
ASA	Active surface area
AST	Accelerated stress testing
BET	Brunauer-emmett-teller
CB-IS	Cross beam (electron impact) ion source
CE	Counter electrode
CEEC	Chemical-electrical energy conversion
CO	Carbon monoxide
COR	Carbon oxidation reaction
CPU	Central processing unit
CV	Cyclic voltammetry/cyclic voltammogram
DAC	Digital-analogue-converter
DAQ	Data acquisition
DDE	Dynamic data exchange
DEMS	Differential electrochemical mass spectrometry
DMFC	Direct methanol fuel cell
DMSO	Dimethyl sulphoxide
ECSA	Electrochemical surface area
EOSA	Electrochemically oxdisable surface area
EQCM	Electrochemical quartz crystal microbalance
FC	Fuel cell
FTIR	Fourier transform infrared
GC	Glassy carbon
GUI	Graphical user interface
$HClO_4$	Perchloric acid
HER	Hydrogen evolution reaction
HOR	Hydrogen oxidation reaction
H_2SO_4	Sulphuric acid
HPLC	High-performance liquid chromatography

HSAC	High surface area carbon
ICP-AES	Inductively coupled plasma atomic emission spectroscopy
KF	Klein flange
KOH	Potassium hydroxide
LSV	Linear sweep voltammogram
LT	Lifetime testing
MEA	Membrane electrode assembly
MID	Mass ion detection
MIMS	Membrane introduction mass spectrometry
MOR	Methanol oxidation reaction
MSCV	Mass spectrometric cyclic voltammogram
MWE	Multiple working electrode
NID	Negative impedance device
OER	Oxygen evolution reaction
ORR	Oxygen reduction reaction
PC	Propylene carbonate
PCTFE	Polychlorotrifluroethylene
PEM	Polymer electrolyte membrane
PEMFC	Polymer electrolyte membrane fuel cell
PTFE	Polytetrafluoroethylene
PVDF	Polyvinylidene Fluoride
QMA	Quadrupole mass analyser
QMS	Quadrupole mass spectrometer
RE	Reference electrode
RDE	Rotating disk electrode
RDS	Rate determining step
RHE	Reversible hydrogen electrode
SCE	Standard calomel electrode
SCSI	Small computer system interface
SDEMS	Scanning DEMS
SEM	Secondary electron multiplier
SHE	Standard hydrogen electrode
SPAIRS	Single potential alteration infrared reflectance spectroscopy
SWE	Single working electrode
TOC	Total organic carbon
TEM	Transmission electron microscopy
VI	Virtual instrument
WE	Working electrode
XPS	X-ray photoelectron spectroscopy
XRD	X-ray diffraction spectroscopy

Abstract

Electrochemical half-cell studies on industrial electrocatalysts contribute significantly towards our understanding of fuel cell processes. However, the study of complex, often overlapping reactions using standard methods is limited to the interpretation of a single electrode current. Presented here are details of the design, construction and characterisation of a differential electrochemical mass spectrometer (DEMS) that enables the in-situ elucidation of electrode currents. The capability of the instrument is demonstrated in two studies. In the first, DEMS is used to resolve the conversion of the methanol oxidation reaction to carbon dioxide on high surface area carbon (HSAC) supported Pt and PtRu catalysts, whilst the second focuses on the corrosion of industrial HSACs, separating partial and complete oxidation processes. Despite that both systems have long since been studied, new insights and understanding can be obtained using DEMS.

Chapter 1
Introduction

At the beginning of a thesis dedicated to the construction of a differential electrochemical mass spectrometer (DEMS), it is essential to first present the broader significance of the instruments intended application in the fundamental study of fuel cell relevant electrochemical reaction processes.

1.1 Background

Rising concerns over climate change and future oil and gas supply security are stimulating a political and economic shift towards a policy of sustainable energy [1]. In order to achieve sustainable energy consumption, aspects of both renewable energy and energy efficiency are required to be developed and encouraged. Sources of renewable energy are, however, not necessarily new concepts with wind-turbines having been present across the landscape for some time. Nevertheless, changing the way in which we produce and supply energy is only one aspect; the energy must also be stored and used efficiently. It is in energy efficiency, therefore, that more efficient chemical-electrical energy conversion (CEEC) devices such as the fuel cell, will play an essential role in replacing today's inefficient energy conversion devices, such as the combustion engine [2]. In contrast to the combustion engine, CEEC devices are not restricted by the Carnot efficiency limitations and can therefore offer higher energy conversion efficiency for the endeavour of sustainable energy [3].

S. J. Ashton, *Design, Construction and Research Application of a Differential Electrochemical Mass Spectrometer (DEMS)*, Springer Theses,
DOI: 10.1007/978-3-642-30550-4_1,

1.1.1 Fuel Cell Technology

The fuel cell CEEC device essentially converts chemical energy (contained in a fuel and oxidant) directly into electrical energy, and crucially does not involve any combustion. In principle, the device operates in a similar manner to an everyday CEEC, the battery; however, in contrast to the battery in which a limited quantity of the chemical energy source is stored internally, a fuel cell uses an external source of fuel and oxidant. An overview of the principle operation of a proton exchange membrane fuel cell (PEMFC) is depicted in Fig. 1.1.

The PEMFC consists of two platinum (Pt) electrodes: an anode and cathode which are separated by a proton exchange membrane (PEM) barrier. The device takes an external source of fuel and oxidant, in this case hydrogen (H_2) and oxygen (O_2), and generates electricity by electrocatalytically converting them into water. This process is a result of two electrochemical reactions which occur at each of the electrodes. At the anode electrode, the hydrogen is oxidised to produce protons and electrons:

$$\text{Anode}: \quad H_2 \rightarrow 2H^+ + 2e^- \quad \mathrm{E}^+ = 0.0\ \mathrm{V_{RHE}}$$

Once generated, both protons and electrons travel from the anode to the cathode electrode; however, they must do so via two different paths imposed by the properties of the PEM. On one hand, the PEM is ion conducting and therefore the protons created at the anode simply pass through the membrane barrier to the cathode. On the other, the PEM is an electrical insulating barrier which means that the electrons must travel to the cathode via an alternate electron conducting pathway through which the current is harnessed. The protons and electrons are then consumed at the cathode electrode in the reduction of oxygen in the following reaction:

$$\text{Cathode}: \quad \frac{1}{2}O_2 + 2H^+ + 2e^- \rightarrow H_2O \quad \mathrm{E}^- = 1.23\ \mathrm{V_{RHE}}$$

The only emission from the hydrogen fuelled PEMFC, therefore, is water vapour. The efficiency of the conversion of chemical into electrical energy in a fuel cell is determined by the potential difference between the anode and cathode electrode potentials. Theoretically, the maximum is 1.23 V determined by the difference between the thermodynamic half-cell potentials (i.e. $\mathrm{E}^- - \mathrm{E}^+$). Unfortunately, kinetic constraints for both anode and in particular, the cathode electrochemical reactions make it impossible for the thermodynamic reversible potential to be realised and large electrode overpotentials result, decreasing the energy conversion efficiency of the fuel cell CEEC device, as depicted in Fig. 1.2.

In order to achieve acceptable energy conversion efficiency (i.e. improve the cell potential), a significant quantity of expensive Pt electrocatalyst must be incorporated into the fuel cell membrane electrode assembly (MEA) to reduce the electrode overpotentials, particularly for the oxygen reduction reaction (ORR) at the cathode. In addition to that, is an extra quantity of Pt electrocatalyst required to ensure that the fuel cell also performs acceptably over its intended lifetime because

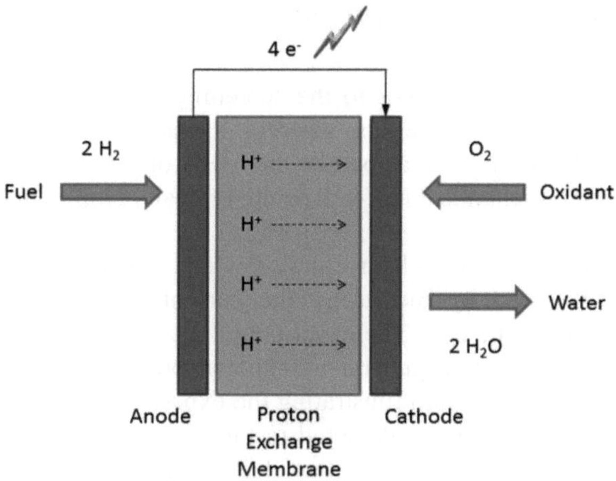

Fig. 1.1 A diagram illustrating the principles behind operation of the hydrogen fed PEMFC.The CEEC device possesses two electrodes that are separated via a proton exchange membrane, and converts an external supply of fuel (H_2) at the anode, and oxidant (O_2) at the cathode, into electricity and water (H_2O)

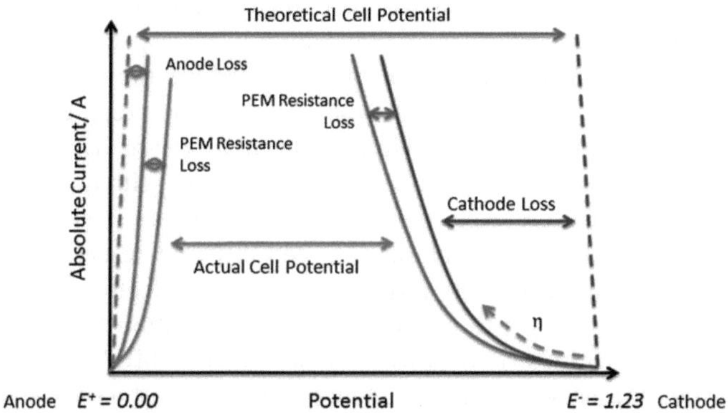

Fig. 1.2 A graph highlighting the origins of PEMFC performance losses. The practical cell potential (ca. 0.6 V) is significantly lower than the theoretical cell potential ($\sim$1.23 V) resulting in a loss in the energy conversion efficiency. A significant proportion of these losses arise from cathode due to the significant overpotential (η) required for the oxygen reduction reaction on Pt

the Pt electrocatalyst performance is not static but unfortunately deteriorates through use [4, 5]. Although adding Pt electrocatalyst can indeed provide an efficient and reliable PEMFC device, the net result of the significant Pt content required is a rather expensive and non-commercially viable CEEC device. Consequently, there is more than a strong incentive to develop electrocatalyst materials that are more active, more durable and less expensive than today's state-of-the-art Pt based fuel cell electrocatalysts [6, 7].

1.1.2 Electrocatalyst Development

There are two feasible approaches to the screening and characterisation of prospective PEMFC electrode materials, whereby the performance of an electrocatalyst is either evaluated in-situ, as part of the MEA of a fuel cell; or ex-situ, in a three–electrode electrochemical half-cell setup, typically employing a rotating disc electrode (RDE) [7].

In the in situ approach, the electrocatalyst is fabricated into the MEA, and its performance is assessed by measuring the current–voltage characteristics (or polarisation curves) of the cell. The exact potentials of the anode and cathode in such measurements in are in general not known precisely, only their potential difference. A simplified diagram illustrating the experimental parameters involved in such an experimental setup is depicted in Fig. 1.3.

In order to assess the performance of an electrocatalysts in an MEA, meticulous control over the MEA fabrication and experimental variables such as fuel and oxidant pressure, flow rate, humidity, and fuel cell operating temperature is required. The cell performance of a particular MEA is typically often published, together with details of the operating temperature, reactant pressure, stoichiometry, MEA preparation, and cell construction etc. Whilst there is no doubt that the assessment of the performance of an electrocatalyst in the fuel cell environment is the ultimate test an electrocatalyst materials performance; controlling every aspect of the MEA fabrication and experimental parameters in order to deduce the performance of the electrocatalyst alone is far from trivial. Not only does the preparation of an MEA require great effort in the research environment, but there are also numerous MEA preparation methods which are often performed by-hand, with questionable reproducibility [8]. It is therefore even more difficult to compare in situ electrocatalyst performance measurement observations between different research groups, and it has been suggested that the incorporation of a prospective electrocatalyst in an MEA should perhaps rather be left to experts in MEA fabrication [9]. The substantial effort, cost and difficulties associated with fabricating MEAs and reproducing experimental conditions with relatively slow accumulation of data, suggests that the in situ assessment of electrocatalyst performance alone is ambiguous, and neither an intuitive nor cost effective approach to the screening and characterisation of prospective fuel cell electrocatalyst materials.

The second approach concerning fuel cell electrocatalyst development are ex-situ methods that typically involve fundamental research using an electrochemical half-cell in aqueous electrolyte, such as the three-electrode RDE setup depicted in Fig. 1.4. The three-electrode electrochemical half-cell setup consists of a working electrode (WE), counter electrode (CE) and reference electrode (RE), each of which are submerged in a conducting aqueous electrolyte. In this setup, the potential of the WE is controlled relative to the RE using a potentiostat. The RE is itself an electrochemical half-cell, whose potential remains at a constant, fixed value. The potentiostat therefore can be used to precisely manipulate the potential of the WE electrode relative to the RE using electroanalytical techniques such as

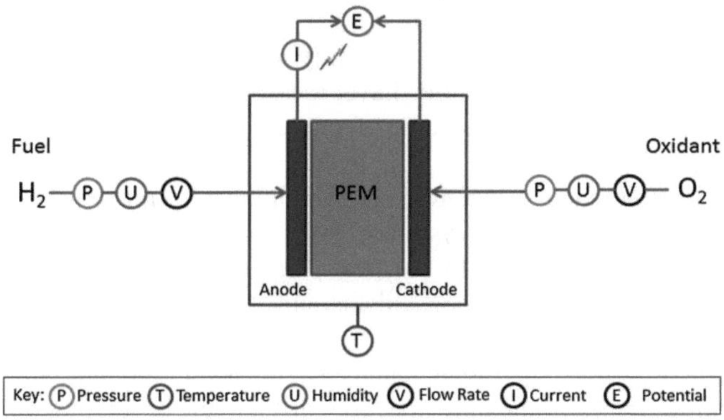

Fig. 1.3 An outline of the experimental setup and the numerous parameters that must be controlled during an in situ MEA experiment for the characterisation of electrocatalyst materials. The electrocatalyst must be fabricated into an MEA and the current is measured as a function of cell potential. These types of experiments are rather complicated and assessing the performance of the electrocatalyst alone is far from trivial

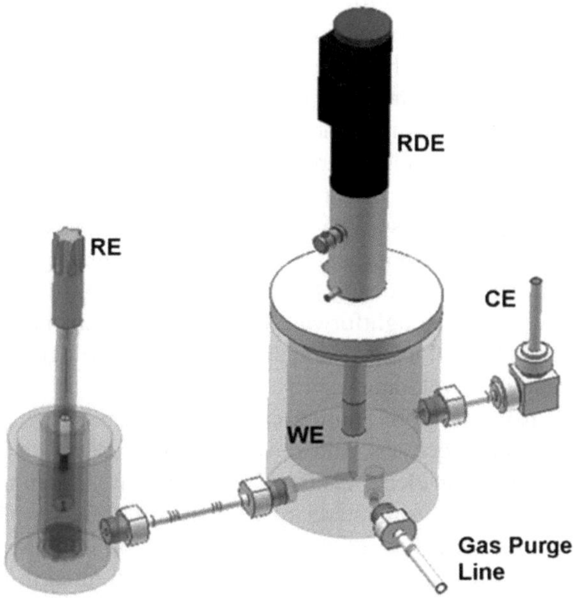

Fig. 1.4 An illustration of an electrochemical half-cell RDE setup that may be employed to characterise the performance of a PEMFC electrocatalyst material ex-situ. The electrocatalyst material is simply deposited on a WE substrate (i.e. no MEA fabrication is required). The setup is rather straightforward compared to in situ methodologies with fewer experimental variables, and the current and WE potential relationship may be measured precisely through the use of a RE

cyclic voltammetry (CV) to study electrochemical processes. The CE meanwhile is required to balance the current flowing through the WE (no current flows through the RE).

The performance and characteristics of electrocatalyst materials can then be accurately evaluated by measuring the current and WE potential relationship of the electro catalytic reaction. The principle advantage of the ex-situ three-electrode electrochemical half-cell over the in situ MEA approach, in terms of electrocatalyst development, is essentially its capability to investigate model electrocatalyst materials. These experimental setups have, for example, been utilised in a number of fundamental studies with particular focus on the electrocatalysis of fuel cell reactions, typically the ORR [7, 10], hydrogen oxidation reaction (HOR) and methanol oxidation reaction (MOR) [11], which have been extensively studied on model electrodes such as polycrystalline [10], single crystal [12] and well-characterised alloy [11, 13–16] Pt surfaces, in order to gain a fundamental understanding of the reaction mechanisms and the relationship between catalyst structure and activity.

Fundamental studies on model catalysts, however, are not always entirely applicable to the more immediate concerns of fuel cell development, although in combination with theory [17] they do provide insights into how the ideal fuel cell electrocatalyst might be designed [11, 18, 19]. A more applied method, therefore, is to perform fundamental research studies in an electrochemical half-cell which directly approach problems associated with fuel cell development needs, such as the characterisation of the activity [7, 20–23], and durability [24, 25] of industrial fuel cell catalysts. In comparison to in situ methods, there are fewer experimental variables to control in the electrochemical half-cell, no MEA fabrication is required and the potential of the WE can be controlled and measured precisely. Consequently, the experiments are comparatively straightforward and cost effective, allowing the more rapid accumulation of experimental data. Electrochemical half-cell setups may therefore be considered as a rather intuitive methodology for effective fuel cell electrocatalyst material screening and characterisation, as well as for fundamental research studies on fuel cell relevant electrochemical processes. This is of course achieved at the expense of the electrocatalyst material being characterised in an aqueous rather than humid gas phase environment of the fuel cell, and consequently the extrapolation of ex-situ observations to the MEA is not necessarily straightforward. A rather complementary approach is therefore crucial toward developing characterisation methodologies to accelerate fuel cell electrocatalyst development.

It is the more applied ex-situ approach to the fundamental study of fuel cell relevant processes, and development of electrocatalyst characterisation methodologies in an electrochemical half-cell that is taken in our laboratory. Nevertheless, such studies using standard RDE setups are not without their own limitations and ambiguities. For example, the study of complex and frequently overlapping electrochemical reactions is in principle limited to the clouded interpretation of a single electrode current. Consequently, more sophisticated techniques that allow the in situ elucidation of electrochemical reactants, intermediates and products, in an electrochemical half-cell such as DEMS are an invaluable research tool.

1.2 Outline and Objectives

The objective of this dissertation was to essentially design, construct and characterise a DEMS instrument, intended for fundamental research studies on fuel cell relevant electrochemical reaction processes. The thesis is organised in the almost chronological order of the design, construction and development processes, followed by two demonstrations of the DEMS application in research.

In order to achieve this goal it was important to first acquire an overview of the DEMS literature, and evaluate the various DEMS design solutions that have appeared previously. A review of the DEMS literature, with particular focus on previous electrochemical cell design solutions is presented in Chap. 2.

In Chap. 3, an overview of the experimental setup along with the design of the critical components of the DEMS instrument built as part of this work is then presented. This section is extended to included necessary details of the hardware and software that were required, or indeed developed to improve the operation and usability of the DEMS instrument.

Once a working DEMS instrument was established, it was crucial to then understand the practical aspects of the instrument and its operation, which is covered in Chap. 4. This includes the evaluation of the effect of experimental variables and a guide for optimising and calibrating the DEMS instrument.

In a first study, the research capabilities of the DEMS instrument is demonstrated in Chap. 5 that revisits the MOR on HSAC supported Pt and PtRu catalysts. Using DEMS, we are able to observe MOR products and evaluate the conversion of methanol to CO_2 as a function of the WE potential using a variety of electroanalytical techniques.

Finally, in Chap. 6 the electrochemical oxidation of various HSAC supports is examined and is a topic which has received relatively little attention using DEMS. In this study, we are able to elucidate electrode currents that arise from partial and complete oxidation of the HSAC supports, as well as the oxygen evolution reaction that occurs on the polycrystalline Au substrate or supported Pt particles.

References

1. Boyle R et al. (2008) Global trends in sustainable energy investment 2008. United Nations Environment Programme
2. Commision E(2003) EUR 20719 EN—Hydrogen energy and fuel cells—a vision of our future
3. Ro ST, Sohn JL (2007) Some issues on performance analysis of fuel cells in thermodynamic point of view. J Power Sources 167(2): 295–301
4. Haas H.R, Davis MT (2009) Electrode and catalyst durability requirements in automotive PEM applications: technology status of a recent MEA design and next generation challenges. ECS Trans 25(1):1623–1631
5. Schmittinger W, Vahidi A (2008) A review of the main parameters influencing long-term performance and durability of PEM fuel cells. J Power Sources 180(1):1–14

6. Gasteiger HA, Marković NM (2009) Just a dream—or future reality? Science 324(5923):48–49
7. Gasteiger HA et al (2005) Activity benchmarks and requirements for Pt, Pt-alloy, and non-Pt oxygen reduction catalysts for PEMFCs. Appl Catal B 56(1–2):9–35
8. Lindermeir A et al (2004) On the question of MEA preparation for DMFCs. J Power Sources 129(2):180–187
9. Gasteiger H, Mathias M (2002) Fundamental research and development challenges in polymer electrolyte fuel cell technology. In: 202nd Meeting of the ECS. Salt Lake City
10. Wiberg GKH, MayrhoferK JJ, Arenz M (2010) Investigation of the oxygen reduction activity on silver—a rotating disc electrode study. Fuel Cells 10(4):575–581
11. Gasteiger HA et al (1993) Methanol electrooxidation on well-characterized platinum-ruthenium bulk alloys. J Phys Chem 97(46):12020–12029
12. Markovic NM et al (1994) Structural effects in electrocatalysis—oxygen reduction on platinum low-index single-crystal surfaces in perchloric-acid solutions. J Electroanal Chem 377(1–2):249–259
13. Stamenkovic VR et al (2007) Trends in electrocatalysis on extended and nanoscale Pt-bimetallic alloy surfaces. Nat Mater 6(3):241–247
14. Markovic NM et al (1995) Electrooxidation mechanisms of methanol and formic-acid on PT-RU alloy surfaces. Electrochim Acta 40(1):91–98
15. Gasteiger H et al (1994) Temperature-dependent methanol electrooxidation on well-characterized PT-RU alloys. J Electrochem Soc 141(7):1795–1803
16. Gasteiger HA et al (1994) Electrooxidation of small organic-molecules on well-characterized PT-RU alloys. Electrochim Acta 39(11–12):1825–1832
17. Koper MTM (2005) Combining experiment and theory for understanding electrocatalysis. J Electroanal Chem 574(2):375–386
18. Markovic NM, Ross PN (2002) Surface science studies of model fuel cell electrocatalysts. Surf Sci Rep 45(4–6):117–229
19. Markovic NM, Ross PN (2000) Electrocatalysts by design: from the tailored surface to a commercial catalyst. Electrochim Acta 45(25–26):4101–4115
20. Mayrhofer KJJ et al (2008) Measurement of oxygen reduction activities via the rotating disc electrode method: from Pt model surfaces to carbon-supported high surface area catalysts. Electrochim Acta 53(7):3181–3188
21. Paulus UA et al (2002) Oxygen reduction on high surface area Pt-based alloy catalysts in comparison to well defined smooth bulk alloy electrodes. Electrochim Acta 47(22–23): 3787–3798
22. Paulus UA et al (2001) Oxygen reduction on a high-surface area Pt/Vulcan carbon catalyst: a thin-film rotating ring-disk electrode study. J Electroanal Chem 495(2):134–145
23. Schmidt TJ et al (1998) Characterization of high-surface area electrocatalysts using a rotating disk electrode configuration. J Electrochem Soc 145(7):2354–2358
24. Mayrhofer KJJ et al (2008) Fuel cell catalyst degradation on the nanoscale. Electrochem Commun 10(8):1144–1147
25. Mayrhofer KJJ et al (2008) Non-destructive transmission electron microscopy study of catalyst degradation under electrochemical treatment. J Power Sources 185(2):734–739

Chapter 2
Differential Electrochemical Mass Spectrometry

The intention of this chapter is to provide an overview of the DEMS technique, the instrument designs and example research applications. This involves the presentation and discussion of previous design solutions, which are separated into three parts: the electrochemical cell, membrane interface and the vacuum system of the mass spectrometer. Example experiments from the DEMS literature are then briefly described which highlight where a particular DEMS electrochemical cell design may be favoured. Ultimately, the appropriate design solution is determined by the intended research application of the DEMS instrument.

2.1 Principle of Operation

Differential Electrochemical Mass Spectrometry (DEMS) is essentially an analytical technique that combines electrochemical half-cell experimentation with mass spectrometry. This allows the in situ, mass resolved observation of gaseous or volatile electrochemical reactants, reaction intermediates and/or products. By correlating the faradaic electrode current with relevant mass ion currents, ambiguous electrochemical reaction processes can be elucidated, whereas standard electrochemical techniques such as the RDE are otherwise limited to the interpretation of a single electrode current. An overview of the DEMS instrument is depicted in Fig. 2.1.

The instrument essentially consists of three crucial components: an electrochemical half-cell, a PTFE membrane interface, and a vacuum system including the quadrupole mass spectrometer (QMS). The role of the electrochemical half-cell is to enable controlled electrochemical experimentation at a WE of interest, and to allow reaction products to be transported to the membrane interface. This interface consists of a microporous PTFE membrane that partitions the aqueous electrolyte of the electrochemical cell from the high vacuum conditions required

S. J. Ashton, *Design, Construction and Research Application of a Differential Electrochemical Mass Spectrometer (DEMS)*, Springer Theses,
DOI: 10.1007/978-3-642-30550-4_2, © Springer-Verlag Berlin Heidelberg 2012

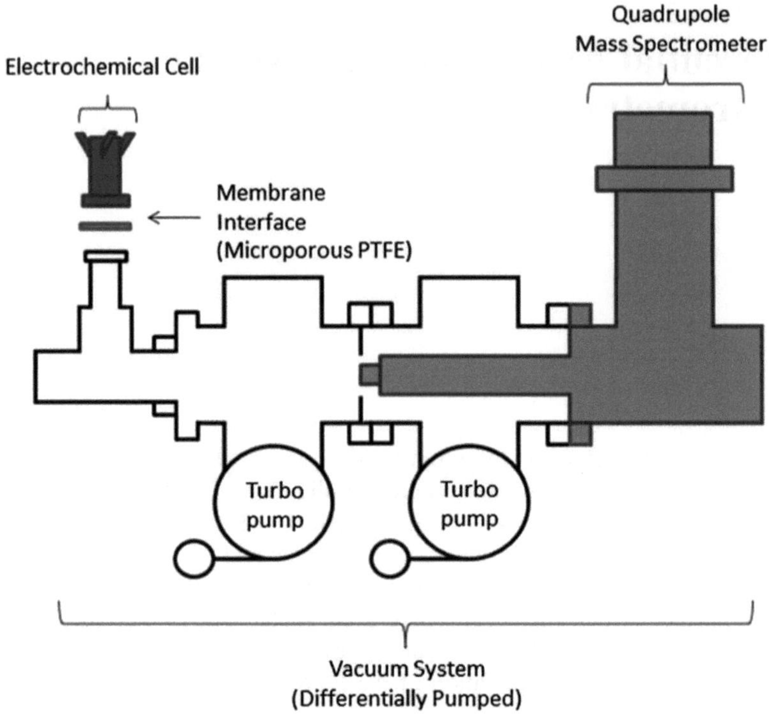

Fig. 2.1 An illustration of the typical components of a DEMS instrument. The DEMS instrument consists of an electrochemical cell that is interfaced to a (2-stage) differentially pumped vacuum system containing a QMS via a microporous PTFE membrane interface

by the QMS. The hydrophobic nature of the membrane prevents the passage of aqueous electrolyte, whilst allowing dissolved gaseous, volatile and relatively non-polar species to permeate, evaporating into the vacuum system. The permeating species may be subsequently observed online by monitoring the relevant mass ion current with QMS. The operation of the QMS, however, requires high vacuum conditions and consequently a specially designed, differentially pumped vacuum system is typically necessary.

The DEMS technique was first established by Wolter and Heitbaum in 1984 [1], however, the origins of utilising mass spectrometry to study electrochemical processes via a PTFE interface may be traced back further to Gadde and Bruckenstein in 1971 [2]. The 'differential' term was chosen in order to distinguish between the time- and potential- resolved correlation of the mass ion and electrode current technique [1] from un-resolved integrating methods [2, 3]. The study of dissolved gases in aqueous solution using mass spectrometry via a membrane interface, however, maybe traced back even further to Hoch and Kok in 1962 [4]. Because DEMS incorporates a membrane interface, the technique is closely related to the analytical technique referred to as membrane introduction mass spectrometry (MIMS). Essentially, the only features that distinguish DEMS from MIMS are the incorporation of an electrochemical half-cell, and the exclusive use of a PTFE microporous membrane interface in DEMS. Despite the almost 30 year history of DEMS, however, no commercial instrument is available and as a result DEMS has been largely limited to the academic research environment where it has been used in a range of fundamental research studies on a variety of different electrodes, albeit by a relatively small number of research groups. Consequently, a number of DEMS design solutions have appeared in the literature over the years, with particular variations in the electrochemical cell design.

2.2 Instrument Design Solutions

The following section is intended to briefly introduce, discuss and evaluate the varieties of DEMS electrochemical cells, membrane interfaces and vacuum system designs that are present in the scientific literature.

2.2.1 Electrochemical Cells

The most flexible aspect of the DEMS instrument is the electrochemical cell design. A number of different DEMS cell designs have gradually evolved over the past 25 or so years, each tailored towards the study of different electrochemical reactions on various types and size of electrodes, ranging from porous [5] or smooth model electrodes [6] to HSAC precious metal catalysts [7] and even fuel cell MEAs [8]. Some of the more noteworthy cell designs will be presented in this section, along with a discussion of their respective advantages and limitations.

2.2.1.1 The Classic Cell

The DEMS technique was first demonstrated using the 'classic' DEMS cell [1]. The most defining feature of this cell is the porous membrane electrode whereby powdered electrode material is deposited directly onto the PTFE membrane interface to form the WE. An illustration of the classic DEMS cell construction is given in Fig. 2.2.

The classic DEMS cell construction consists of a modified vacuum flange containing a stainless steel frit that mechanically supports the porous membrane WE. The membrane electrodes were originally prepared by painting a lacquer containing small metallic particles (e.g. Pt) directly onto the PTFE membrane [9]; however, this was soon adapted so that an electrocatalyst layer could be sputter deposited directly onto the PTFE membrane interface [5], presumably to provide a more homogeneous layer. The electrochemical cell and membrane electrode are

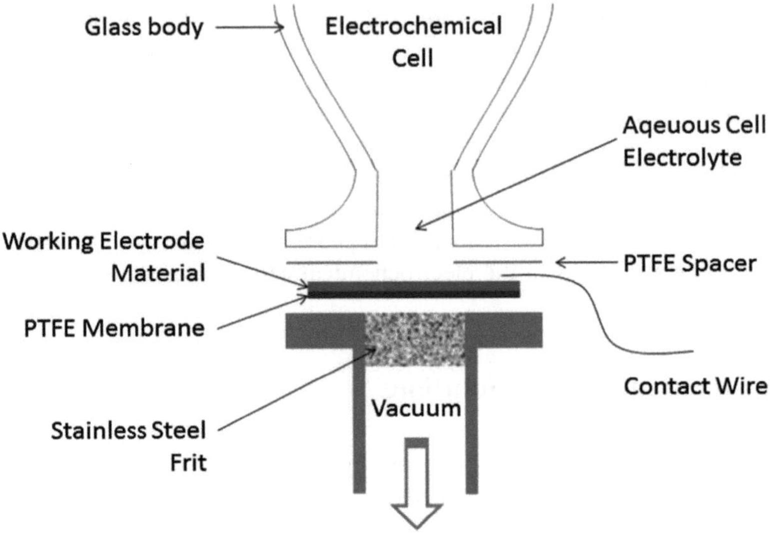

Fig. 2.2 An illustration of the 'classic' DEMS cell construction. The WE material is deposited directly onto the microporous PTFE membrane interface. The CE and RE (not shown) are then positioned in the electrolyte above the membrane WE

sealed between the vacuum flange and a PTFE ring which is compressed by the body of the electrochemical cell compartment. The electrochemical cell body has been typically constructed of glass although a PTFE cell has also been demonstrated [10]. Electrical contact to the WE meanwhile is achieved using a platinum wire, and the CE and RE are typically located within the single electrochemical cell compartment, positioned above the porous membrane WE via the multiple connections situated in the cell hood. These connections also serve to allow electrolyte to be deaerated, whilst also allowing electrolyte to be pumped through the cell if desired.

The principle advantage of this classic DEMS cell design is a fast mass spectrometry response time, typically <0.1 s and high collection efficiency of ca. 0.5 for lacquer and 0.9 for sputtered electrode [9]. Both are a consequence of the microscopic transport distances between the electrode surface and the vacuum side of the membrane. The cell may also be optimised to possess a small volume (ca. 1 to 2 ml) with static electrolyte, allowing very economical use of costly isotope labelled reactants for studies of electrochemical mechanisms [9]. The extremely close proximity of electrode material to the membrane interface does, however, introduce certain experimental difficulties. For instance, when using volatile reactants that readily evaporate through the membrane, their concentration at the electrode surface can be significantly reduced compared to the bulk, even with electrolyte convection. The competing process of evaporation through the membrane and adsorption of reactants onto the electrode surface could also lead to a shift in reaction equilibrium [9]. Moreover, whilst convection of electrolyte over the WE may be achieved by positioning the cell electrolyte inlet hosing close to the membrane electrode (perhaps to some extent analogous to a wall-jet configuration), the true mass transport of reactants is very likely to be difficult to define and control. This is problematic when studying either steady state or diffusion limited processes, although an adaptation of the classic cell design incorporating a rotating rod to induce convection did appear to resolve this issue [11]. The design is further limited to experimentation with materials that can be deposited onto the PTFE membrane interface i.e. it is therefore impossible to, for example, study technical or smooth bulk electrodes such as single crystal or even polycrystalline electrode surfaces. Furthermore, the characterisation of WE material sample loading or utilisation and homogeneity (which is crucial in electrocatalysis) on the non-conducting membrane is difficult to quantify. Finally, the RE is also positioned within the electrochemical cell which can introduce impurities that can have dramatic consequences on the electrochemistry observed [12].

The trends in electrocatalyst research since the early inception of DEMS led focus away from powdered electrode materials that the classic DEMS cell was originally designed for, to more model systems such as polycrystalline and single crystal electrodes. As a consequence of these limitations, few new research studies utilising this DEMS cell alone have appeared for the past 15 years (with the exceptions of Ref. [13, 14]) and hence its designation in this review as the 'classic' DEMS cell.

2.2.1.2 Thin-Layer Cell

The thin-layer design first appeared in 1990 and was initially developed to allow the DEMS study of smooth bulk electrodes i.e. polycrystalline or single crystal electrodes [15]. The DEMS thin-layer flow cell construction illustrated in Fig. 2.3.

In contrast to the classic DEMS cell, the WE is now separated from the membrane interface via a thin-layer of electrolyte allowing considerably more flexibility in terms of the type of electrode that may be studied. Due to the miniature nature of this cell design ($\sim$1 μL volume [15]), the CE and RE are now placed in separate compartments, whose capillaries (which can be connected to either the electrolyte inlet or outlet) lead to the thin-layer WE compartment. However, the electrochemical reaction products must now diffuse from the electrode surface and across the thin layer before it may evaporate through the membrane to be detected by the QMS.

The obvious advantage of this design is improved versatility because a variety of well-characterised, bulk model WEs may be studied, and the preparation of a porous membrane electrode is no longer necessary. Furthermore, the flow of electrolyte and reactants over the electrode may be defined more exactly and easily controlled compared to the classic DEMS cell. The separation of the WE from the membrane interface, however, results in inherently longer delay times ca. 2–3 s [9, 15], although the collection efficiency is rather high $\sim$0.9 [9] under static

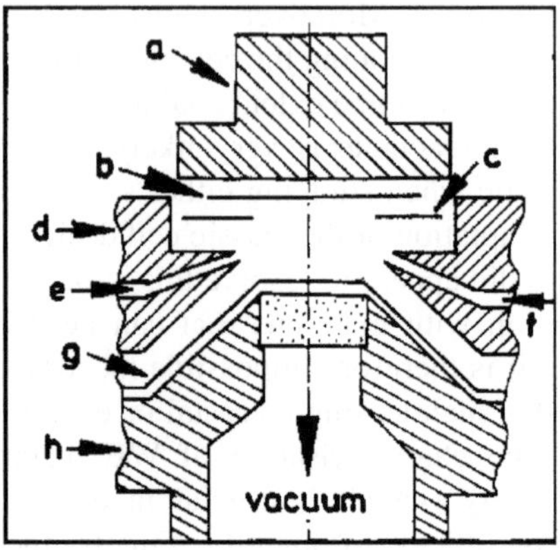

Fig. 2.3 A sketch of the thin-layer flow cell (*a*) electrode support, (*b*) platinum sheet (WE), (*c*) circular PTFE spacer, (*d*) cell body, (*e*) electrolyte outlet and connection to counter electrode, (*f*) electrolyte inlet, (*g*) porous PTFE membrane separating electrolyte and vacuum (support by a porous steel frit), (*h*) connection to vacuum chamber. Not shown, reference electrode connected by a third capillary. The WE in this cell design is separated from the membrane interface via a thin-layer of electrolyte allowing technical WE(s) to be used in DEMS. Reprinted with permission from Ref. [15]. Copyright 1990 American Chemical Society

electrolyte conditions. Under continuous electrolyte flow conditions efficiencies of only 0.2 or below are observed at practical flow rates of at least 1 μl s^{-1} [15]. Whilst the thin-layer cell can still be utilised to determine product formation rates under electrolyte flow conditions, there are a number of serious drawbacks. Firstly, the diffusion of reaction products across the thin layer is competing with the flow of electrolyte through the cell and therefore, any reaction products formed at the electrode close to the cell outlet may not be able to diffuse to the membrane before exiting the cell; the result of which is reduced collection efficiency. A second issue, regarding volatile and gaseous reactants is that these will, in addition to reacting at the WE, also evaporate through the membrane leading to a significant decrease in the concentration gradient along the cell flow field. This DEMS thin-layer cell design is therefore more appropriate for the study of desorption products under static electrolyte flow conditions [16–18]. As a consequence of the relatively poor cell collection efficiency during electrolyte flow conditions, in cases where convection of the WE is desired, the dual thin-layer flow cell is usually preferred.

Overall, the thin-layer cell is an ideal approach to the study of a variety of technical WE under static electrolyte conditions such as stripping or desorption measurements. However, in much electrochemical experimentation it is desirable to have a controlled and continuous flow of electrolyte and reactants over the WE surface for which the thin-layer cell is less appropriate.

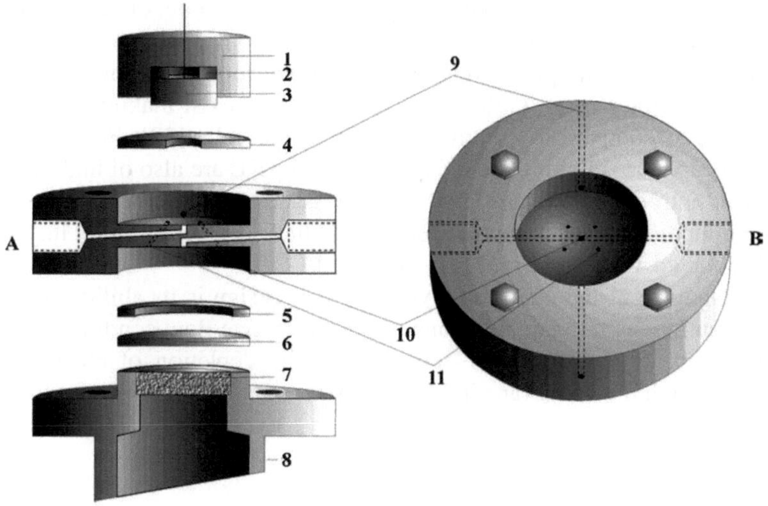

Fig. 2.4 A drawing of the dual thin-layer flow cell (1) Kel-F support, (2) Kalrez, (3) electrode, (4, 5) Teflon spacer, (6) porous Teflon membrane, (7) stainless steel frit, (8) stainless steel connection to MS, (9) capillaries for flushing with Ar, (10) inlet–outlet capillaries, (11) connecting capillaries. (**a**) Cross-section and top-view (**b**) of Kel-F cell body. The separation of the WE and membrane interface into two separate compartments avoid the problems associated with the original thin-layer design during the study of electrochemical reactions under electrolyte flow conditions. Reproduced by permission of ECS—The Electrochemical Society [3]

2.2.1.3 Dual Thin-Layer Flow Cell

A large quantity of the research performed in the past 10 years using DEMS has employed the dual thin-layer flow cell, depicted in Fig. 2.4.

The cell was initially developed to incorporate an electrochemical quartz crystal microbalance (EQCM), for combined DEMS and EQCM studies [3] but has since then, also been used to combine DEMS with attenuated total reflectance fourier transform infrared spectroscopy (ATR-FTIR) [19, 20]. In contrast to the thin-layer flow cell described previously, the WE and membrane are now divided into two separate thin-layer compartments: a WE compartment and a membrane interface sampling compartment. As a result, two key problems associated with the thin-layer DEMS cell are avoided: firstly, the bulk concentration of gaseous and volatile reactants at the WE are no longer depleted by evaporation through the membrane, and secondly the cell collection efficiency is enhanced to ~0.4 under electrolyte flow conditions [9]. This is a consequence of the increased retention time of the electrolyte over the membrane interface. The dual thin layer cell has been typically constructed with either passivated titanium or PCTFE (Kel-f).

The dual-thin layer cell design is better suited to the monitoring of continuous faradic reactions, with controlled hydrodynamics in the determination of reaction product formation rates and turn over frequencies. A drawback of the dual thin-layer flow cell, however, is a comparably slow response time, often ca. 2 s [21] or more because of the finite time required for the reaction products to be transported from the WE to the sampling compartment. Furthermore, the flow of electrolyte over the WE surface is not uniform and static regions exist making this particular cell difficult to compare to common hydrodynamic cells [22]. Finally, the thin-layers and capillaries linking the CE and WE to the RE are also of high resistance, which can introduce WE potential control issues, and the DEMS setup has required a modified 3-elecrode electrochemical half- cell setup which incorporates two CE coupled via an external resistor [21, 23, 24].

The advantage of the dual thin-layer flow cell design is its ability to allow the study of continuous faradaic reactions under electrolyte flow conditions, avoiding the limitations of the thin-layer cell, namely the depletion of electrochemical reactants through the membrane interface and poor collection efficiency under flow conditions.

2.2.1.4 Capillary Inlet

An alternative approach to positioning the WE close to the membrane interface, is to position a PTFE membrane covered capillary inlet vacuum interface close to WE [25–28]. The capillary inlet (also known as the pinhole inlet) DEMS technique is illustrated in Fig. 2.5.

The inlet consists of a capillary whose tip is covered by [25], or imbedded with [26] a microporous PTFE membrane material. The capillary, for example, consists of a 0.6 mm diameter glass tubing with a 0.3 mm inlet diameter which is positioned between 10 and 20 μm away from the electrode surface [26]. Electrochemical reaction products which are formed close to the capillary inlet may then be either sampled by the membrane interface or simply diffuse away into the electrolyte bulk. The quantity of electrochemical reaction species entering the vacuum system is, therefore, inherently smaller than the regular membrane interface. The use of a capillary inlet was also extended in the multi-electrode scanning DEMS (SDEMS) cell allowing localised monitoring of reaction products over an electrode array [27, 28]. In this case the capillary inlet is attached to a three-dimensional piezoelectric-driven positioning system so that it may be moved along the length of a band of electrodes of different type or composition for catalyst screening. Here the capillary inlet consisted of a PTFE capillary tube, 0.15 mm in diameter with 20 nm pore Gore-Tex membrane, and was typically positioned 100 μm away from electrode [27].

The advantages of the capillary inlet system is essentially that DEMS can be used in combination with single crystal electrodes in the hanging meniscus configuration [25, 26] whereas the SDEMS system allows either a large number of electrode samples to be screened within a single experiment, or a particularly large electrode surface to be spatially mapped [27, 28]. The use of a rather small membrane surface area and capillary inlet also offers some advantages in terms of the DEMS vacuum system, which can be of a simpler design because of the

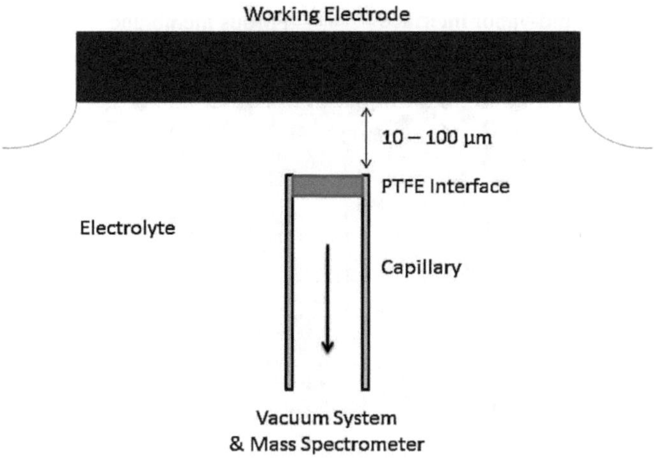

Fig. 2.5 An illustration of the capillary inlet DEMS electrochemical cell. The microporous PTFE membrane is incorporated into the end of a capillary whose interface is then position closely to the WE surface. This allows the study of either (1) single-crystal electrodes in a hanging meniscus configuration, or (2) multiple electrode arrays

inherently lower gas flux passing through the capillary (compared to the membrane of other cells), as will be discussed later in Sect. 2.2.3. There are, however, a number of serious limitations. For instance, only a small fraction of the electrochemical reaction product is sampled resulting in an inherently lower instrument sensitivity. Furthermore, the capillary inlet interface samples from a larger volume than the cylindrical volume between the capillary inlet and electrode surface, and the amount of electrochemical reactants that enter the membrane capillary is sensitive to the capillary inlet position relative to the electrode surface. This perhaps makes calibration and quantitative experimentation somewhat tricky. The superposition of the products planar diffusion away from the surface and the spherical diffusion to the capillary may also lead to complicated time dependence and longer response times [9]. Finally, the capillary inlet would be inappropriate for experimentation requiring convection over the electrode surface.

2.2.2 Membrane Interfaces

In DEMS a microporous PTFE membrane is used to interface the aqueous electrolyte of the electrochemical cell with the high vacuum required by the mass spectrometer. The membrane acts as a selective barrier, preventing the passage of aqueous electrolyte into the vacuum system whilst allowing electrochemical reactions species to evaporate through into the vacuum to be detected by the mass

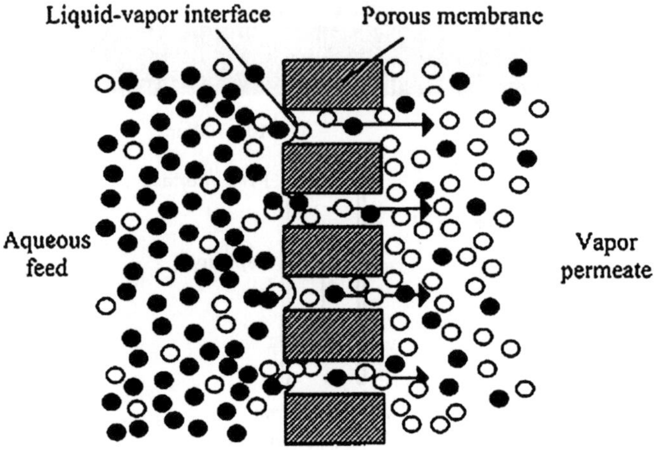

Fig. 2.6 An illustration of the vacuum membrane distillation process through a microporous PTFE membrane. The hydrophobic and microporous nature of the membrane interface prevents the passage of liquid electrolyte whilst allowing dissolved gaseous and/or volatile species to readily evaporate through the membrane pores. Reprinted from Ref. [29], Copyright 2001, with permission from Elsevier

spectrometer. The process is analogous to vacuum membrane distillation, a process which is depicted in Fig. 2.6.

The hydrophobic properties of the membrane prevent the penetration of the aqueous solution into the pores, resulting in a vapour-liquid interface at each pore entrance. The passive transport of species through the membrane pores is via the vapour phase and is instigated by the pressure drop across the membrane between the hydrostatic pressure of the aqueous electrolyte and the vacuum system of the mass spectrometer. The prevention of liquid transport through the hydrophobic PTFE membrane is rather dependent on a critical parameter, specifically the membrane pore size. For aqueous liquid the hydrophobic membrane must possess a critical pore radius of < 0.8 μm in order to prevent liquid phase transport through the membrane [1]. This is a value determined by the Kelvin law and is dependent on the surface tension of the liquid, the contact angle between the liquid and membrane, and the applied pressure. For a given pore size, therefore, a critical hydrostatic penetration pressure exists above which the liquid phase will be transported through the membrane. Hydrophobic membranes with a narrow pore size distribution and high level of porosity are therefore preferred for this application [30]. The membrane specifications found in the DEMS literature, when provided, are summarised in Table 2.1.

In practice, the PTFE microporous membranes generally used in DEMS are typically $\sim$ 0.02 μm in size with a thickness between 50 and 110 μm. By using even smaller membrane pore sizes molecular flow in the pores is ensured, and the influence of isotropic effects during the gas transport are avoided [1, 38]. Almost all studies using DEMS have been performed in aqueous electrolyte with the

Table 2.1 Details of the DEMS microporous PTFE membrane interface specifications contained in the scientific literature

Pore size/μm	Thickness/μm	Porosity/%	Supplier	Geometric Area/cm^2	Literature Source
0.2	60	50	Scimat	–	[7, 19, 21, 31, 32]
0.17	60	50		–	[33–35]
0.02	110	–	Schleicher & Schuell	–	[8, 36]
0.02	50	–	–	1.13	[3]
0.02	75	50	Gore	0.39	[34, 37]

In general the pore dimensions are ≤0.2 μm with thicknesses ≤110 μm

exception being studies in organic electrolytes such as propylcarbonate (PC), dimethylsulfoxide (DMSO) [39] and dimethylcarbonate [13] which prove to be non-wetting to the DEMS microporous PTFE membranes. There are of course also other membrane materials which may be suited to DEMS such as polypropylene and polyvinylidene fluoride (PVDF), although these are unlikely to possess any advantages in consideration of the superior durability, and chemical resistance of PTFE. Microporous PTFE membranes, meanwhile, are generally widely available in flat and capillary forms which can be unsupported; however, the PTFE membrane is more commonly laminated onto a nonwoven polypropylene support web to improve mechanical strength and handling.

The use of a PTFE microporous membrane does impose certain limitations on the type of species that may be detected using DEMS. In order for a species to readily evaporate through a PTFE membrane, it must be either gaseous or volatile, and relatively non-polar [9]. Although the closely related MIMS technique may employ a wide variety of membrane material, DEMS has so far exclusively used a microporous PTFE membrane interface because of a number of its properties, namely its robust and reliable hydrophobic character, mechanical stability, durability and chemical resistance in the electrochemical environment.

Despite the non-wetting properties of the PTFE microporous membrane, a significant quantity of aqueous electrolyte is still able to evaporate through the membrane pores and into the vacuum system owing to the significant pressure gradient across the membrane. It is therefore crucial that a reasonably sized vacuum system can sufficiently pump the continuous flux of gas penetrating the membrane and achieve the operating pressures required by the QMS. This parameter, however, is not a typical specification of PTFE membranes which are for instance, more commonly employed for filtration in HPLC or woven into high performance clothing. There are some literature values for the total gas flux value of a PTFE membrane, which was calculated to be 0.09 mbar l s^{-1} at 20 °C assuming a water vapour pressure of 23 mbar, and 0.4 mbar l s^{-1} at elevated temperature of 40 °C [9]. Values other than these are rather scarce, and presumably depend on the exact specifications of the membrane i.e. pore size and thickness.

Finally, a membrane is not the only method of interfacing an electrochemical cell with a mass spectrometer. For example, a capillary had been used to sample the gas above a closed electrochemical cell [40] or thin-layer flow cells have been combined with electrospray mass spectrometry [41–43] and inductively coupled plasma atomic emission spectroscopy (ICP-AES) for trace metal analysis [44]. The techniques, however, do not offer the time and potential resolved capabilities of DEMS.

2.2.3 Vacuum Systems & Mass Spectrometer

Despite that the hydrophobicity and pore size of the microporous PTFE membrane prevents the transport of liquid water, a considerable quantity is still able to evaporate through the membrane pores into the vacuum system. Consequently, a steady flux of gas $\sim$0.09 mbar l s^{-1} cm^{-2} that largely consists of water vapour (in addition to gaseous, volatile and non-polar species) enters the vacuum system [9]. The requirement of the vacuum system is to therefore pump at such a rate, that the necessary operating pressure of the mass spectrometer is obtained, typically <1.0 E^{-5} mbar. The exact design of the vacuum system can have important implications on the performance of the DEMS instrument, both in terms of pumping efficiency and mass spectrometer sensitivity [21]. However, the details of the DEMS vacuum system such as the exact operating pressures or membrane gas flux are seldom presented or discussed in the literature, and the vacuum system is often simply described as being *differentially pumped.*

Differentially pumped vacuum systems are not specific to DEMS but are generally useful when requiring to measure trace constituents of gas at relatively high pressure, for example 1 mbar, using a mass spectrometer that must operate at $<10^{-5}$ mbar vacuum pressure. The differentially pumped vacuum system essentially consists of two or more, individually pumped vacuum chambers that are connected by a small aperture or skimmer. To illustrate why a single vacuum system is not sufficient, we can consider a reasonably sized vacuum system with a pumping speed of 300 l s^{-1} which must operate at 10^{-5} mbar pressure. The total gas flux that can be pumped by this system is 3×10^{-3} mbar l s^{-1}. In order to interface this vacuum system to an electrochemical cell via a microporous PTFE membrane with a total gas flux of 0.09 mbar l s^{-1} cm^{-2} then the acceptable membrane interface geometrical area is 0.033 cm^{-2}. Although a single pumping stage may be sufficient for the capillary inlet DEMS vacuum system [26], such a small membrane surface area is rather impractical for the classic and thin-layer DEMS cells, which employ somewhat larger geometrical areas $\sim$1 cm^2[3, 15]. In order to allow a larger membrane interface geometric area and therefore gas flux, a differentially pumped vacuum system must be employed. By using a vacuum system with two chambers, and allowing a relatively high vacuum pressure of $\sim 10^{-3}$ mbar in the first pumping chamber, with a pumping speed of ca. 90 l s^{-1} a total gas flux of 0.09 mbar l s^{-1} cm^{-2} can be pumped, allowing a more practical membrane geometric area of $\sim$1 cm^2. The pressure in the 2nd chamber, containing the QMS, therefore, may be determined by controlling the conductance of gas between the 1st and 2nd pumping chamber using an aperture, in order to obtain the necessary operating pressure of the mass spectrometer $<10^{-5}$ mbar. For example, a 200 l s^{-1} pumping speed in the second chamber would allow an operating pressure of the mass spectrometer of 10^{-5} mbar to be achieved using an aperture with a conductance of ~ 2 mbar l s^{-1}. A circular aperture with a radius of 0.85 cm would therefore be acceptable, and can be easily machined.

Previous DEMS instruments vacuum systems have employed a number of different types of pumps for the 1st chamber, such as a rotary vane pump [8, 45], diffusion pump [11], or turbo pump [9]; however, all instruments used a turbo pump for the second pumping stage containing the mass spectrometer. It has been claimed that a higher sensitivity of the DEMS achieved by positioning the ion source between the first and second DEMS chambers [21] although exact details are not given. This can presumably be attributed to allowing a higher pressure at the ion source, which increases the number of ions produced relative to the vacuum system background. Once a differentially pumped vacuum system has been constructed and can accept a certain gas flux, the vacuum system is not necessarily limited to DEMS, as highlighted in a previous instrument [45], whereby the DEMS vacuum system and QMS was used in combination with thermo-gravimetric analysis. For the related MIMS technique, however, differentially pumped vacuum systems are generally not required because total gas fluxes through the membranes are much smaller owing to the different non-porous membrane types used [34].

2.3 Research Applications

The purpose of this section is to briefly analyse the types of experiments that may be performed using DEMS, and highlight when certain DEMS electrochemical cells are favoured.

2.3.1 Radio-Isotope Labelled Experimentation

The very first DEMS study using radio-isotope labelled reactants was simply entitled 'Does the oxide layer take part in the oxygen evolution reaction on Platinum' [46] and aimed to elucidate whether or not the PtO layer formed on Pt electrode >0.9 V_{RHE} was involved in the oxygen evolution at potentials >1.5 V_{RHE} according to the following reaction:

$$2H_2O + PtO_{(bulk)} \rightarrow O_2 + 4H^+ + 4e^- + PtO_{(bulk)}$$

This study nicely demonstrates the power of DEMS using radio-isotope labelled reactants, employing the classic DEMS cell. Although Pt electrodes possess relatively poor activity toward the oxidation of water, it is considered to be a model electrode useful for fundamental studies on the oxygen evolution reaction (OER) mechanism, and there was some debate concerning the role played by the Pt oxide layer. By using $H_2^{18}O$ to form an oxide layer of $Pt^{18}O$ species, the DEMS instrument could be used to observe whether the O_2 evolved during the oxidation of $H_2^{16}O$ contained any $^{16}O^{18}O$ species in circumstances where the oxide layer does take part, or if only $^{16}O^{16}O$ species are formed. In this study, only $^{16}O^{16}O$ species were observed, indicating that the oxide is not consumed during the OER. In this case the classic DEMS cell was quite suitable because the experiment purely concerned the fundamental study of a surface reaction and does not require the use of more technical bulk electrodes, and that the cell could be optimised to possess a small volume using little of the expensive radioisotope labelled water.

2.3.2 *Characterisation of Organic Adsorbates*

The first application of the thin-layer DEMS cell, which allowed the use of smooth electrodes by separating the WE from the membrane interface, was to study the cathodic desorption and hydrogenation of benzene to cyclohexane on an annealed Pt electrode, in 0.5 mol dm^{-3} H_2SO_4 [15]. The general procedure for this type of study is described by the following process:

$$\begin{matrix}\text{Electrolyte} \\ + \\ \text{Adsorbate}\end{matrix} \xrightarrow{\text{adsorption}} \begin{matrix}\text{Electrolyte} \\ \text{Exchange}\end{matrix} \xrightarrow{\text{desorption}} \begin{matrix}\text{Electrolyte} \\ + \\ \text{Desorbate}\end{matrix}$$

These types of experiments involve adsorbing a certain reaction species, such as benzene at 0.5 V_{RHE}, which is saturated in the starting cell electrolyte. The electrolyte is then exchanged with the pure electrolyte removing the saturated species, whilst maintaining the WE potential at a fixed value. The WE potential is then swept to a potential where the pre-adsorbed species may desorb from the WE surface. In this study, it was observed that benzene partially desorbs without reaction in the $H_{upd/ads}$ potential region, however, at more negative potentials where the hydrogen evolution reaction takes place on Pt, the benzene is hydrogenated and desorbs in the form of cyclohexane [15]. In adsorption and then desorption measurement such as these, where no convection of the electrolyte is required during the electrochemical experimentation (except to exchange electrolytes), the thin-layer cell is very suitable and can incorporate almost all types of technical electrode. Finally, in addition to benzene there are numerous other desorption studies such as acetone [47], bicyclic aromatics [48, 49], ethene [6, 16] systems, each employing the thin-layer cell.

2.3.3 *Study of the Electro-oxidation of Small Organic Compounds*

One of the first ever DEMS studies focused its attention towards the study of the methanol oxidation reaction (MOR) which is of fundamental research interest for direct methanol fuel cell (DMFC) development [37]. Since then, the MOR [7, 21, 31, 50, 51] and other potential organic fuels that are oxidised to carbon dioxide, such as ethanol [32, 52–54] have been quite extensively studied using DEMS on a number of electrode systems. The overall MOR may be described by the following reaction formula:

$$CH_3OH + H_2O \xrightarrow{E} CO_2 + 6H^+ + 6e^-$$

However, it was long since suspected that the MOR does not reach completion based on strange dependences of the MOR current on the convection of electrolyte using conventional RDE techniques [55]. This assumption could, however, not be confirmed using standard RDE techniques alone because only a single faradaic electrode current is observed. In contrast, DEMS is able to monitor in situ both the consumption of methanol [56], and the generation of partially oxidised intermediates [57] and carbon dioxide [7, 21, 31, 50, 51] allowing the calculation of the MOR current efficiencies and study of the reaction mechanism. The dual thin-layer flow cell is particularly suitable for the study of the MOR under flow conditions owing to its advantages over the thin-layer flow cell during experiments using electrolyte convection i.e. there is less depletion of reactant in the electrolyte along the flow vector, and higher electrode collection efficiencies are achieved.

2.4 Conclusions

The intention of this chapter was to provide an overview of the literature concerning DEMS, with particular focus on previous design solutions and their respective advantages and limitations, prior to the design of the DEMS instrument to be constructed in this thesis. The DEMS instrument requires the custom design and fabrication of three parts: the electrochemical half-cell, membrane interface and vacuum system.

Various DEMS electrochemical cell designs have appeared over the years with varying degrees of success. In recent years, few publications have appeared using the classic DEMS electrochemical cell that deposited the electrode material directly onto the membrane interface, primarily because electrochemists today are rather more interested in the study of smooth electrode surfaces, which cannot be incorporated with the microporous PTFE membrane and must therefore be separated. The most implemented DEMS cell designs of the past 10 years or more, are the thin-layer, and in particular, the dual thin-layer electrochemical flow cells. In these designs the electrode and membrane are separated via either a thin-layer layer of electrolyte, or by placing them in two separate thin-layer compartments. These cells allow the use of bulk electrodes with sufficient sensitivity, whereby the thin-layer cell is favoured under static electrolyte conditions, whilst the dual-thin layer cell is more suited to continuous flow experiments. In general, the requirements of the electrochemical cell is to incorporate the electrode and vacuum interface of choice, allow controlled electrochemical experimentation at the WE, transport electrochemical products to the vacuum interface and obtain sufficient time and potential resolution.

All DEMS design solutions employ a microporous PTFE membrane in order to interface the aqueous electrolyte of the electrochemical cell and the vacuum system of the mass spectrometer. The membrane permeation process is analogous to vacuum membrane distillation. In order to prevent the passage of liquid water, the radius of the membrane pores must be $<0.08\ \mu m$. The use of a PTFE membrane does, however, impose certain restrictions on the electrolytes that may be used, along with the electrochemical reaction species that can be detected, because only dissolved gaseous, volatile and relatively non-polar species readily pass through the microporous membrane. Despite the hydrophobic properties of the PTFE membrane, a considerable amount of water still evaporates into the vacuum system, often requiring a specifically designed differentially pumped vacuum system.

The requirement of the DEMS vacuum system is to pump away the gas flux that passes through the PTFE membrane at such a rate that the necessary vacuum conditions required for the mass spectrometer are maintained. In order for this to be achieved using both a sensibly sized vacuum system, and a reasonable membrane interface surface area (or gas flux), a two-stage differentially pumped vacuum system is often employed. Such vacuum systems are not specific to DEMS but are often used when a gas at relatively high pressure ~ 1 bar must be measured online using mass spectrometry. Details of the total gas flux through the PTFE membrane, vacuum system and corresponding operating pressures of a DEMS instrument are, however, rather scarce in the literature.

DEMS has been utilised to study a variety of electrochemical reactions on various electrode systems. For instance, the classic DEMS has been used in combination with radio-isotope electrochemical reactants to elucidate the origins of electrochemical surface species. The development of the thin-layer cell for DEMS allowed the study of smooth electrodes and is particularly suitable to the study of electrode adsorption potential induced desorption products which do not require convection. However, convection can be desirable for experiments involving continuous faradaic reactions, such as the electrochemical oxidation of small organic compounds. In such cases the dual thin-layer flow cell is preferred.

Overall, there is no commercially available DEMS instrument and the design of the custom parts must be tailored toward the intended research application of the instrument.

References

1. Wolter O, J Heitbaum (1984) Differential electrochemical mass-spectroscopy (Dems) - a new method for the study of electrode processes. berichte der bunsen-gesellschaft-physical chemistry Chem Phys 88(1):2–6
2. Bruckenstein S, RR Gadde (1971) Use of a porous electrode for in situ mass spectrometric determination of volatile electrode reaction products. J Am Chem Soc 93(3):793–794
3. Jusys Z, H. Massong, H. Baltruschat (1999) A new approach for simultaneous DEMS and EQCM: electro-oxidation of adsorbed CO on Pt and Pt-Ru. J Electrochem Soc 146(3):1093–1098
4. Hoch G, B Kok (1963) A mass spectrometer inlet system for sampling gases dissolved in liquid phases. Arch Biochem Biophys 101(1):160–170
5. Willsau J, J Heitbaum (1984) The influence of Pt-activation on the corrosion of carbon in gas-diffusion electrodes - a dems study. J Electroanal Chem 161(1):93–101
6. Schmiemann U, H Baltruschat (1992) The adsorption of ethene at pt single-crystal electrodes - desorption products and observation of multiple adsorption states BY DEMS. J Electroanal Chem 340(1–2):357–363
7. Jusys Z, J Kaiser, RJ Behm (2002) Composition and activity of high surface area PtRu catalysts towards adsorbed CO and methanol electrooxidation - A DEMS study. Electrochimica Acta 47(22–23):3693–3706
8. Seiler T et al. Poisoning of PtRu/C catalysts in the anode of a direct methanol fuel cell: a DEMS study. Electrochimica Acta 49(22–23):3927–3936
9. Baltruschat H (2004) Differential electrochemical mass spectrometry. J Am Soc Mass Spectrom 15(12):1693–1706
10. Wieckowski A (1999) Interfacial electrochemistry: theory: experiment, and applications. Marcel Dekker, New York
11. Fujihira M, T Noguchi (1993) A novel differential electrochemical mass-spectrometer (dems) with a stationary gas-permeable electrode in a rotational flow produced by a rotating rod. J Electroanal Chem 347(1–2):457–463
12. Mayrhofer KJJ et al. (2009) An electrochemical cell configuration incorporating an ion conducting membrane separator between reference and working electrode. Int J Electrochem Sci 4(1):1–8
13. Lanz M, P Novak (2001) DEMS study of gas evolution at thick graphite electrodes for lithium-ion batteries: the effect of gamma-butyrolactone. J Power Sources 102(1-2):277–282
14. Novak P et al. (2000) Advanced in situ methods for the characterization of practical electrodes in lithium-ion batteries. J Power Sources 90(1):52–58
15. Hartung T, H Baltruschat (1990) Differential electrochemical mass-spectrometry using smooth electrodes - adsorption and h/d-exchange reactions of benzene on pt. Langmuir 6(5):953–957
16. Loffler T, H Baltruschat (2003) Temperature dependent formation of multiple adsorption states from ethene at polycrystalline Pt and Pt(111) electrodes studied by differential electrochemical mass spectrometry. J ElectroanalChem 554:333–344
17. Sanabria-Chinchilla J et al. (2006) A DEMS study of the electrocatalytic hydrogenation and oxidation of p-dihydroxybenzene at polycrystalline and monocrystalline platinum electrodes. J Appl Electrochem36(11):1253–1260
18. Sanabria-Chinchilla J et al. (2007) Electrocatalytic hydrogenation and oxidation of aromatic compounds studied by DEMS: Benzene and p-dihydroxybenzene at ultrathin Pd films electrodeposited on Au(hkl) surfaces. J Colloid Interface Sci 314(1):152–159
19. Heinen M et al.(2007) In situ ATR-FTIRS coupled with on-line DEMS under controlled mass transport conditions—A novel tool for electrocatalytic reaction studies. Electrochimica Acta 52(18):5634–5643
20. Vielstich W, Gasteiger H, Yokokawa H (2009) Handbook of fuel cells: advanced in electrocatalysis, materials, diagnostics and durability, vol. 5. Wiley, New York

21. Jusys Z, Behm RJ (2001) Methanol oxidation on a carbon-supported pt fuel cell catalysta kinetic and mechanistic study by differential electrochemical mass spectrometry. J Phys Chem B 105(44):10874–10883
22. Holzbecher E (2006) Thin layer flow cell modelling. in COMSOL users conference,Frankfurt
23. Colmenares LC et al.(2009) Model study on the stability of carbon support materials under polymer electrolyte fuel cell cathode operation conditions. J Power Sources 190(1):14–24
24. Ming-fang L et al. (2010) Oxidation of carbon supports at fuel cell cathodes: differential electrochemical mass spectrometric study. Chin J Chem Phys 23(4):442
25. Gao YZ, et al. (1994) New online mass-spectrometer system designed for platinum single-crystal electrode and electroreduction of acetylene. J Electroanal Chem 372(1–2):195–200
26. Wonders AH et al. (2006) On-line mass spectrometry system for measurements at single-crystal electrodes in hanging meniscus configuration. J Appl Electrochem 36(11):1215–1221
27. Jambunathan K, Jayaraman S, Hillier AC (2004) A multielectrode electrochemical and scanning differential electrochemical mass spectrometry study of methanol oxidation on electrodeposited ptxruy. Langmuir 20(5):1856–1863
28. Jambunathan K, Hillier AC(2003) Measuring electrocatalytic activity on a local scale with scanning differential electrochemical mass spectrometry. J Electrochem Soc 150(6):E312–E320
29. Urtiaga AM et al.(2001) Parallelism and differences of pervaporation and vacuum membrane distillation in the removal of VOCs from aqueous streams. Sep Purif Technol 22-23:327–337
30. Drioli E, Calabro V, Wu Y(1986) Microporous membranes in membrane distillation. Pure Appl Chem 58(12):1657–1662
31. Jusys Z, Kaiser J, Behm RJ (2003) Methanol electrooxidation over pt/c fuel cell catalysts: dependence of product yields on catalyst loading. Langmuir 19(17):6759–6769
32. Wang H, Jusys Z, Behm RJ (2006) Ethanol electro-oxidation on carbon-supported Pt, PtRu and Pt3Sn catalysts: A quantitative DEMS study. J Power Sources 154(2):351–359
33. Schmidt TJ et al. (2001) On the CO tolerance of novel colloidal PdAu/carbon electrocatalysts. Journal of Electroanalytical Chemistry 501(1-2):132–140
34. Smith SPE, Casado-Rivera E, Abruna HD (2003) Application of differential electrochemical mass spectrometry to the electrocatalytic oxidation of formic acid at a modified Bi/Pt electrode surface. J Solid State Electrochem 7(9):582–587
35. Pastor E, Schmidt VM (1995) Electrochemical reactions of ethene on polycrystalline au electrodes in acid-solution studied by differential electrochemical mass-spectrometry and isotope labeling. J Electroanal Chem383(1-2):175–180
36. Rosca V, Beltramo GL, Koper MTM (2004) Hydroxylamine electrochemistry at polycrystalline platinum in acidic media: a voltammetric, DEMS and FTIR study. J Electroanal Chem 566(1):53–62
37. Willsau J Wolter O, Heitbaum J (1985) On the nature of the adsorbate during methanol oxidation at platinum—a dems study. J Electroanal Chem 185(1):63–170
38. Jusys Z Liaukonis J, Vaskelis A(1991) The catalytic-oxidation of hypophosphite on nickel studied by electrochemical mass-spectrometry. J Electroanal Chem 307(1–2):87–97
39. Eggert G, Heitbaum J (1986) Electrochemical reactions of propylenecarbonate and electrolytes solved therein—a dems study. Electrochimica Acta 31(11):1443–1448
40. Hibbert D.B, Churchill CR (1984) Kinetics of the electrochemical evolution of isotopically enriched gases. Part 3-Hydrogen and deuterium evolution on platinum and platinised tungsten trioxide. J Chem Soc, Faraday Trans 1: Phys Chem Condens Phases 80(7):1977–1984
41. Diehl G, Karst U (2002) On-line electrochemistry—MS and related techniques. Anal Bioanal Chem 373(6):390–398
42. Deng H, Berkel GJV (1999) A thin-layer electrochemical flow cell coupled on-line with electrospray-mass spectrometry for the study of biological redox reactions. Electroanalysis 11(12):857–865
43. Bökman CF, Zettersten C, Nyholm L (2004) A setup for the coupling of a thin-layer electrochemical flow cell to electrospray mass spectrometry. Anal Chem 76(7):2017–2024

44. Zhou F (2005) Electrochemistry combined on-line with atomic mass spectrometry and related techniques for trace-metal analysis and electrode-reaction studies. TrAC Trends Anal Chem 24(3):218–227
45. Wasmus S Samms SR, Savinell RF (1995) Multipurpose electrochemical mass-spectrometry—a new powerful extension of differential electrochemical mass-spectrometry. J Electrochem Soc 142(4):1183–1189
46. Willsau J, Wolter O, Heitbaum J (1985) Does the oxide layer take part in the oxygen evolution reaction on platinum—a dems study. J Electroanal Chem 195(2):299–306
47. Bansch B. et al. (1989) Reduction and oxidation of adsorbed acetone at platinum-electrodes studied by dems. J Electroanal Chem 259(1–2):207–215
48. Loffler T et al.(2003) Adsorption and desorption reactions of bicyclic aromatic compounds at polycrystalline and Pt(111) studied by DEMS. J Electroanal Chem 550:81–92
49. Vrestal J et al. (1999) Mass spectrometric study of desorption and hydrogenation of biphenyl, naphthalene and t-butylbenzene on poly- and monocrystalline platinum electrodes. J Electroanal Chem 461(1–2):90–93
50. Jusys Z et al. (2002) Activity of PtRuMeOx (Me = W, Mo or V) catalysts towards methanol oxidation and their characterization. J Power Source 105(2):297–304
51. Wang H, Baltruschat H (2007) DEMS study on methanol oxidation at poly- and monocrystalline platinum electrodes: the effect of anion, temperature, surface structure, ru adatom, and potential.J Phys Chem C 111(19):7038–7048
52. Fujiwara N, Friedrich KA, Stimming U (1999) Ethanol oxidation on PtRu electrodes studied by differential electrochemical mass spectrometry. J Electroanal Chem 472(2):120–125
53. Jiang L et al. (2007) Ethanol electrooxidation on novel carbon supported Pt/SnOx/C catalysts with varied Pt : Sn ratio. Electrochimica Acta 53(2):377–389
54. Colmenares L et al. (2006) Ethanol oxidation on novel, carbon supported Pt alloy catalysts—model studies under defined diffusion conditions. Electrochimica Acta 52(1):221–233
55. Bagotzky VS, Vassiliev YB, Khazova OA (1977) Generalized scheme of chemisorption, electrooxidation and electroreduction of simple organic-compounds on platinum group metals. J Electroanal Chem 81(2):229–238
56. Wang H, Loffler T, Baltruschat H (2001) Formation of intermediates during methanol oxidation: a quantitative DEMS study. J Appl Electrochem 31(7):759–765
57. Anastasijevic NA, Baltruschat H, Heitbaum J (1989)DEMS As a tool for the investigation of dynamic processes—galvanostatic formic-acid oxidation on a pt electrode. J Electroanal Chem 272(1–2):89–100

Chapter 3
Design and Construction of the DEMS Instrument

In this chapter, details of the DEMS instrument design and construction created as part of this thesis are presented. The instrument employs a dual thin-layer electrochemical flow cell and microporous PTFE membrane interfaced to a high vacuum system containing a QMS. The vacuum system design possesses a tubular aperture to control and direct the flux of gas through the cross-beam ion source, whilst a 3-stage differentially pumped vacuum construction provided optimum operating pressures of the QMS in order to maximise instrument sensitivity. The actual functionality and usability of the DEMS instrument, however, also relies on additional instrumentation, control and data acquisition hardware and software. This includes the development of custom hardware such as the negative impedance device (NID), and custom Labview programmed electrochemical DEMS CV, Macro and QMS calibration software, along with a DEMS data analysis program. Overall, the design presented here provided a very capable, reliable and usable DEMS instrument.

3.1 Design and Development Process

When starting the design and development of any construction, a number of basic technical requirements must first be considered in order to provide a specific direction in which the project should proceed. In the construction of an instrument such as the DEMS, the intended application of the scientific apparatus was first defined: the key objective of this DEMS instrument was essentially to allow the in situ observation of electrochemical reaction species, enabling the study and elucidation of electrochemical processes relevant to fuel cell technologies on both model and applied electrocatalysts. There were of course more general requirements considered such as reliability, functionality, performance, life-time and even price. Once these requirements were outlined (where possible) initial design

S. J. Ashton, *Design, Construction and Research Application of a Differential Electrochemical Mass Spectrometer (DEMS)*, Springer Theses,
DOI: 10.1007/978-3-642-30550-4_3,

concepts were then thought-out, although in the case of the DEMS instrument the concept of interfacing an electrochemical cell with a mass spectrometer to study electrochemical reactions via a membrane interface has existed for some time. There were therefore, details of DEMS design solutions contained within the scientific literature, an overview of which was presented and discussed previously in Chap. 2. These of course aided considerably in the design and development of this instrument, and I am also extremely thankful to Dr. Jusys for our friendly discussions concerning DEMS, which offered an early insight into issues that must be considered during the design and development process. Once a design was finalised, a development phase followed in which the cycle of fabrication, evaluation and when necessary redesign or modification was pursued until a satisfactory instrument construction was achieved.

This chapter, which concerns the design and construction of the DEMS instrument begins with an overview of the whole DEMS instrument. The attention is then focused on the more detailed discussion of the designs of the parts more specific to DEMS, namely the electrochemical cell, membrane interface and vacuum system of the QMS. However, rather than beginning each section with a long laborious list of design requirement that are neither pleasant to read or indeed write, the designs themselves are simply presented with a discussion of the important features and critical aspects that were considered. The focus then shifts onto the necessary hardware and software required to operate the DEMS instrument. This largely includes commercial hardware for the electrochemistry, vacuum system and QMS, however, a custom NID was also developed to perform *iR*-compensation of the WE in the DEMS electrochemical half-cell. The majority of the software for the DEMS instrument consists of a number of custom software programs developed in the Labview graphical programming environment, with the exception of the commercial Quadstar 32-bit software required to operate the QMS. These software programs combine with the DEMS hardware in two different measurement setups: the first is for common DEMS experimentation, whereas the second is intended for the calibration of the QMS. These measurement setups are each described, along with the conceptual architecture of the custom Labview software programs. The chapter finally ends with a presentation of the DEMS data analysis software architecture.

3.2 DEMS Instrument Overview

The constituents of the overall DEMS instrument can be divided into two main categories: firstly, there are intrinsic DEMS components i.e. the electrochemical flow cell, membrane interface and vacuum system including QMS; and secondly, there are the instrumentation, control and data acquisition hardware and software that are essential to the operation of the instrument. An overview of the affiliation between the DEMS components, and various hardware and software required for the instrument to function is illustrated in Fig. 3.1.

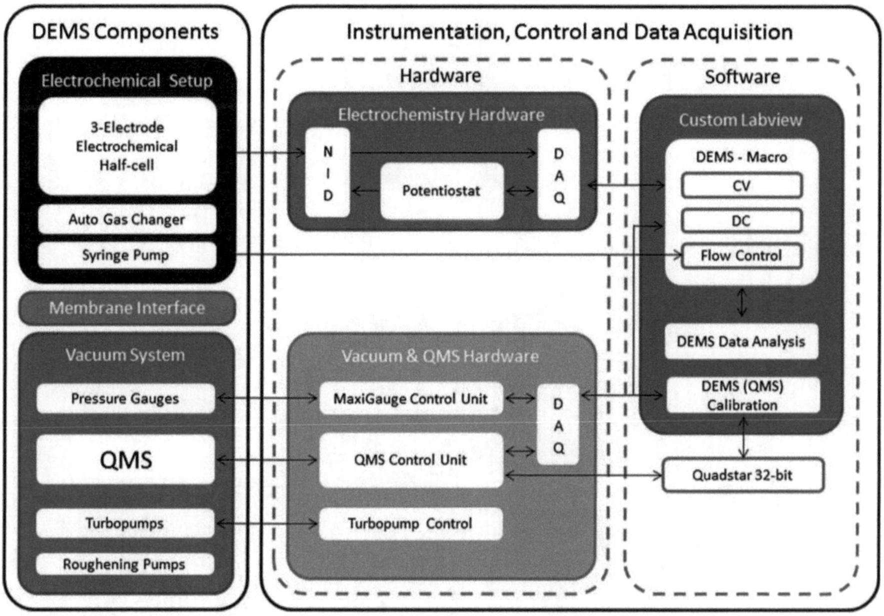

Fig. 3.1 A diagram presenting an overview of the various components contributing to the whole DEMS instrument setup. The instrument consists of the key DEMS components, namely the electrochemical cell, membrane interface and vacuum system including QMS. The operation of the DEMS components of course relies on additional instrumental control and data acquisition hardware and software

Although in DEMS we are essentially combining an electrochemical half-cell with mass spectrometry, they are nonetheless operated and controlled as discrete entities that happen to be (albeit intentionally) interfaced by a membrane. The instrumentation, control and data acquisition hardware is therefore divided up into those that are required to perform and measure the electrochemistry from those required to operate and measure the QMS mass ion currents and vacuum system pressures. The control of this instrumentation along with the acquisition and synchronisation of experimental data from the electrochemical, vacuum and QMS hardware is, however, combined using a number of custom programmed Labview software to significantly enhance the functionality and usability of the DEMS instrument, facilitating its use as a research tool.

An overview of the actual DEMS instrument is depicted in the photograph in Fig. 3.2. The location of the electrochemical flow cell, vacuum system, electrochemistry hardware, and QMS and vacuum system hardware are each highlighted.

The electrochemical flow cell setup is situated on a sturdy aluminium bench top which also (along with the movable trolley) serves to house and support the vacuum system and turbo pump controllers. The rotary vane pumps required to back the turbo pumps are meanwhile located in a separate sound isolation box positioned to the left of the vacuum system. The electrochemistry, QMS and

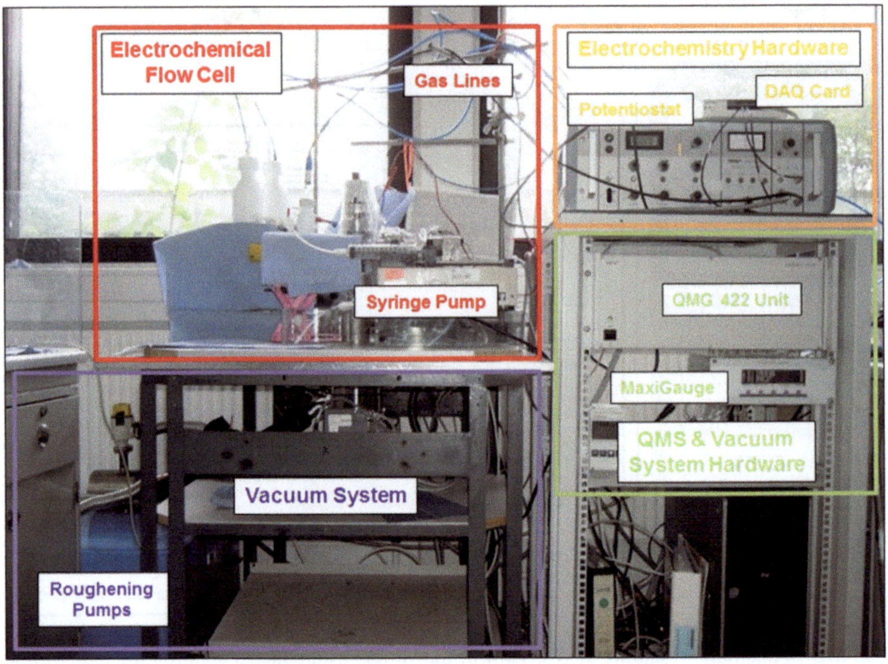

Fig. 3.2 Photograph of the DEMS instrument constructed as part of this thesis. The electrochemical cell setup is located on a sturdy bench top (or trolley) which serves to house and support the vacuum system and related hardware. Supplementary hardware for the electrochemistry, QMS and vacuum pressure measurement are meanwhile located on a separate equipment rack

vacuum pressure control and measurement instrumentation are situated on a separate hardware rack located to right of the vacuum chamber. These are then controlled by the DEMS user via a combination of the custom Labview DEMS software and Quadstar 32-bit software, operating on a Microsoft Windows based desktop computer (not shown in Fig. 3.2).

3.2.1 Principle Components

An overview of the principal instrument components inherent to DEMS, namely the electrochemical cell, membrane interface and vacuum system including QMS, of the instrument designed and constructed in this thesis is illustrated in Fig. 3.3.

The electrochemical cell is of a dual thin-layer flow cell design, incorporating a WE compartment and membrane interface compartment which is attached to the DEMS vacuum system. By employing the cell in a three-electrode electrochemical half-cell setup (not shown in Fig. 3.3) the potential of the WE can be controlled to

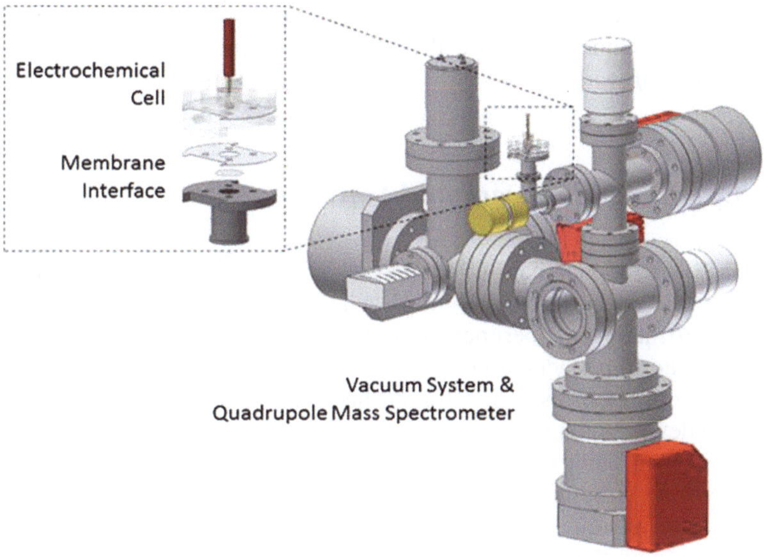

Fig. 3.3 A technical drawing of the DEMS instrument construction highlighting the electrochemical cell, membrane interface and vacuum system including QMS. The electrochemical cell incorporating the WE and membrane interface is fixed to the vacuum flange of a leak valve connecting the vacuum system

study electrochemical reactions. A controlled flow of electrolyte meanwhile transports the electrochemical reaction products away from the WE to the membrane interface where they may be sampled into the vacuum system.

The membrane is used to interface the aqueous electrolyte of the electrochemical cell with the high vacuum conditions of the QMS. In DEMS this is achieved using microporous PTFE membrane which is able to sample electrochemical reaction products in a process that is analogous to vacuum membrane distillation. The membrane prevents the passage of aqueous electrolyte whilst allowing dissolved gaseous, volatile and relatively non-polar species to evaporate into the vacuum system of the QMS.

The DEMS vacuum system is used to achieve the high vacuum conditions required for mass spectrometry. In this DEMS instrument, this was attained using a three-stage differentially pumped vacuum system allowing the total flux of gas that enters the vacuum system from the membrane interface to be pumped at such a rate that the operating conditions of the QMS are maintained. The vacuum system incorporates a commercial Pfeiffer Vacuum QMA 422 quadrupole mass analyser with a 90 ° off-axis SEM and cross-beam electron impact ion source (CB-IS). The QMS allows the relevant mass ion currents of electrochemical reaction species which evaporate through the membrane interface to be monitored in situ with the electrochemistry performed at the WE.

Of course, in order for the electrochemical cell, vacuum system and QMS to function and combine as DEMS, they rely on a number of additional operating hardware and software.

3.2.2 Operating Hardware and Software

The DEMS operational hardware required for the electrochemistry includes a Wenking POS2 potentiostat to control and manipulate the potential of the WE during electrochemical experimentation. Precise control of the WE potential also necessitates the implementation of a custom designed and built NID to achieve *iR*-compensation of the solution (or in this case stabilising) resistance: a rather uncommon device amongst electrochemical half-cell setups. Finally, a commercial precision syringe pump (KDS 200) was employed in order to control the electrolyte flow rate through the DEMS dual thin-layer electrochemical flow cell.

The vacuum system hardware includes the necessary roughening- and turbo-pumps, along with the corresponding turbo pump control units, vacuum pressure gauges including MaxiGauge pressure gauge control and measurement unit. The hardware required by the QMS, meanwhile, includes the QMG 422 control unit, Rf-generator and EP 422 electrometer amplifier.

The QMS is operated using the commercial Quadstar 32-bit software provided with the mass spectrometer. The DEMS instrument experimentation, however, is performed using custom Labview software including CV, DC and Macro software. A custom Labview program was further more designed and developed to allow the straightforward calibration of the QMS. Labview is a graphical programming environment that allows engineers and scientists to develop sophisticated measurement, test and control systems. Finally, in order to aid the analysis of DEMS data, custom Labview data analysis software was developed to perform common data treatment tasks that are specific to DEMS.

3.3 Electrochemical Half-Cell Setup

An illustrated overview of the entire electrochemical flow cell setup is given in Fig. 3.4 that (described from left to right) consists of an electrolyte reservoir(s), CE compartment, the DEMS dual thin-layer flow cell containing WE and membrane interface mounted to the QMS vacuum system, a RE compartment and finally, the syringe pump flow control. .

Each of the electrochemical flow cell components are interconnected using flanged 1.6 mm internal diameter commercial PTFE tubing with UNF 1/4 in. 28 G fittings, typically used in high performance liquid chromatography (HPLC). The electrolyte reservoir(s) is attached via a PTFE 4-way connection which is also attached to the sealed CE compartment and the DEMS flow cell inlet. Note that the

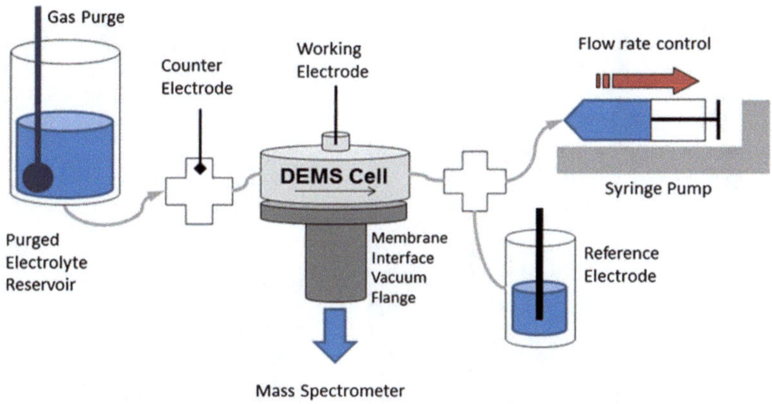

Fig. 3.4 A sketch of the three-electrode DEMS electrochemical flow cell setup. The CE and RE are separated from the WE compartment and positioned in two separate sealed compartments on the cell inlet and outlet respectively. The electrolyte from a purged reservoir is allowed to flow through the cell at a volumetric rate determined by the syringe pump (operating in withdrawal mode) positioned on the cell outlet

remaining connections of the 4-way flange may either be stoppered, connected to a second fluid reservoir or a controlled electrolyte drain line. The DEMS flow cell outlet is then attached to another 4-way connection, linking the sealed RE compartment and the syringe pump, whereby the final connection is stoppered. The purged electrolyte is thus able to flow from the supply reservoir(s), through the DEMS flow cell (over the WE and then membrane interface) and into the syringe at a rate determined by the syringe pumping speed. The syringe pump is operated in withdrawal mode, at flow rates lower than the maximum flow of electrolyte through the cell, as determined by the hydrostatic pressure of the supply reservoir(s) and the size of the flow cell capillaries. The syringe pump thus effectively limits the flow of electrolyte, rather than sucking it through the cell. Images of the actual electrochemical flow cell setup are presented in Fig. 3.5.

It should be noted that the CE compartment does not lie in the path of the electrolyte between the supply reservoir(s) and DEMS flow cell, which is important in preventing electrochemical reaction products (such as H_2 or O_2) that arise from the CE, from contaminating the inlet flow of electrolyte and interfering with the electrochemistry at the WE. For similar reasoning, the saturated calomel electrode (SCE) RE is also located on the exit of the electrochemical flow cell in order to avoid the contaminating effects of chlorides arising from the SCE before the electrolyte contacts the WE [1]. In general, the lengths of all interconnecting PTFE tubing are kept to a minimum with the intention of maintaining the purged state of the electrolyte from the reservoir, as it flows through the cell. All electrochemical flow cell reservoir(s), CE and RE compartments and connecting tubing, meanwhile, consist of PTFE whereas the dual thin-layer flow cell components are Kel-f (PCTFE), both of which are chemically inert to acidic and basic electrolytes that are commonly used in electrochemistry, such as H_2SO_4 and KOH.

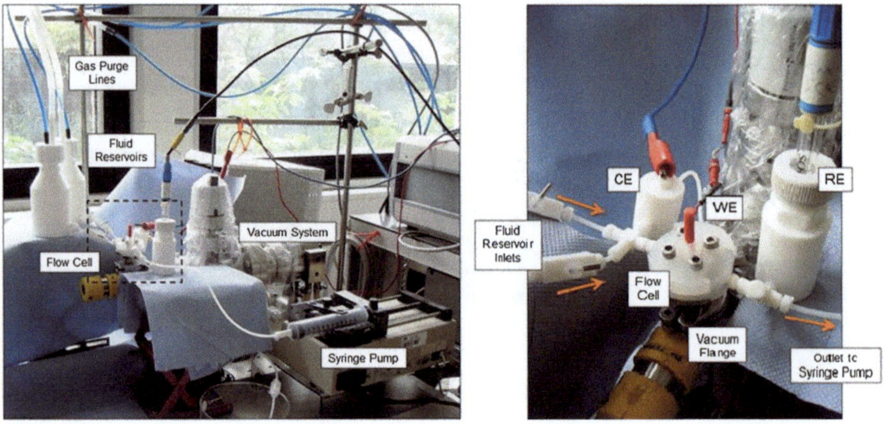

Fig. 3.5 Photographs of the DEMS electrochemical flow cell setup constructed as part of this thesis. All cell components are fabricated from either PTFE or Kel-f (PCTFE), and are connected using 1.6 mm internal diameter PTFE tubing. The electrolyte flow rate is determined by the rate of withdrawal of electrolyte from the gas purged electrolyte reservoirs

3.3.1 Dual Thin-Layer Flow Cell Design

The DEMS electrochemical cell produced in this work is of a modular dual thin-layer flow cell design possessing two small (thin-layer) compartments connected via four narrow capillaries: an electrochemical cell volume containing the WE and membrane interface sampling compartment. An exploded overview of the DEMS electrochemical cell design and membrane interface is presented in Fig. 3.6.

This cell construction can be divided into three principle interlocking components: (1) an electrode block, (2) flow-through block (each fabricated from Kel-f) and (3) a stainless steel KF vacuum connection flange. When these components are combined with the WE (including Kel-f screw piece), two PTFE gaskets and a microporous PTFE membrane; the electrochemical cell and membrane sampling compartments are formed which are partitioned by the flow-through block (5 mm thick). The cell and membrane gaskets are cut from commercial PTFE sheets, and are typically 0.25 mm thick. The gasket thickness and size of the holes cut through the centre of each PTFE gasket form and define the dimensions of the two compartments: (1) the electrochemical cell volume that is enclosed by the WE above, and flow-through block below, and (2) the membrane interface sampling compartment, which is similarly enclosed by the flow-through block above, and microporous PTFE membrane below. The membrane is meanwhile supported by a porous metal stainless steel frit (40 μm pore size, Applied Porous Technologies©) embedded in the centre of the vacuum flange connection, providing mechanical support to the membrane when it is exposed to the vacuum. Finally, each of these components are inserted together, fixed and sealed tight using four M4 bolts that screw directly into the vacuum flange.

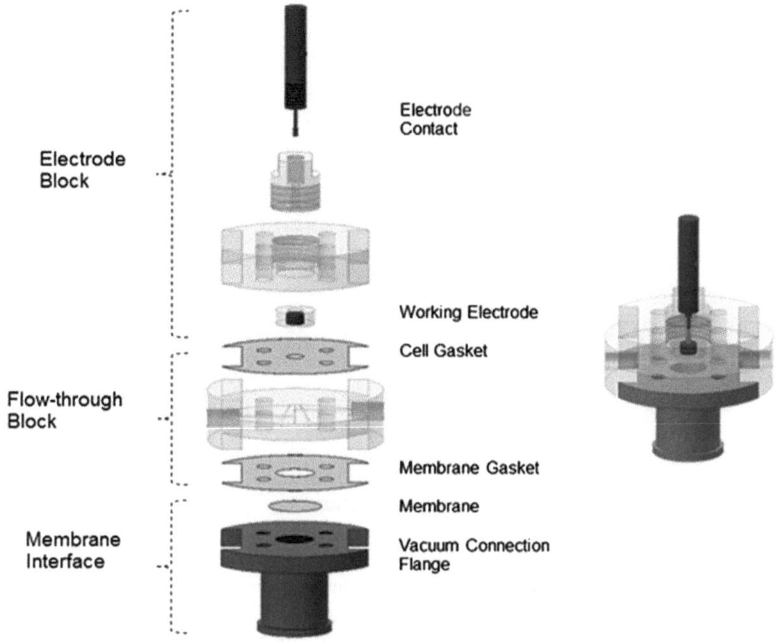

Fig. 3.6 Exploded drawing of the modular DEMS electrochemical flow cell design and membrane interface constructed as part of this thesis. The cell components lock together and are tightened and sealed using four M4 bolts which screw into the vacuum flange. The WE and membrane interface compartment volumes are determined by the thickness of the PTFE gaskets

The modular design presented here is intended to allow a considerable degree of flexibility and/or modification to the DEMS electrochemical cell parameters. Using this modular system, individual cell components can be modified, exchanged or replaced when necessary, without requiring a completely new cell construction. This is particularly beneficial in the cycle of construction, evaluation, modification and when necessary re-design or modification for a new research application, because it reduces the number of parts that are required to be fabricated minimising, time, labour and material costs. More importantly, this modular construction allows for the optimisation of cell parameters, for instance, the PTFE cell gaskets can be exchanged with gaskets of different dimensions to optimise cell volumes, or an alternative flow-through block could be employed to use a different electrolyte flow regime over the WE surface or membrane interface. In the long-term, parts which may deteriorate more quickly in time may also be more easily replaced.

In regards to the electrolyte flow through the cell, the exact flow regimes over the WE and membrane interface is determined by both the arrangement of the inlet and exit capillaries of the flow-through block, and the shape of the punctured holes of the WE compartment gasket. In this thesis, two dual-thin layer designs were constructed: the dual-cyclone, and dual-linear thin-layer flow cell.

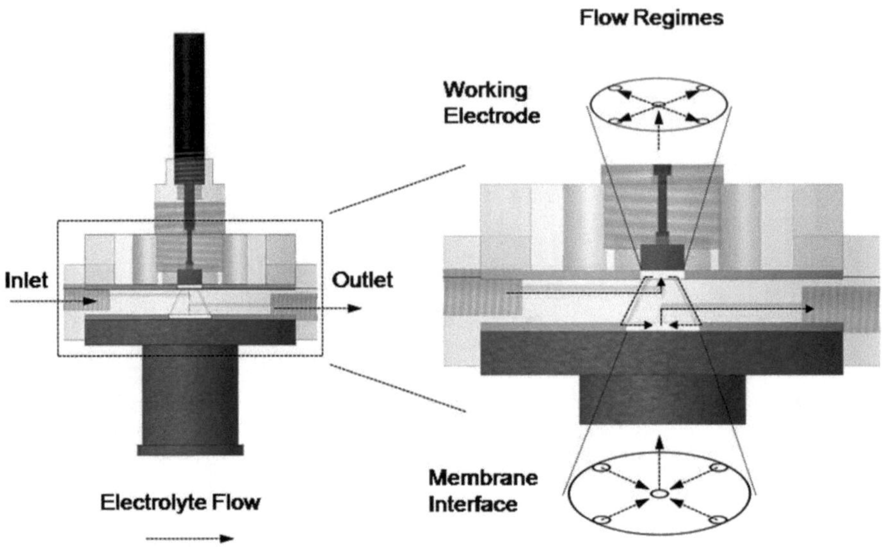

Fig. 3.7 Drawing of the electrolyte flow regime through the dual-cyclone thin-layer electrochemical flow cell. The electrolyte flows from the centre of the WE outwards to four capillaries that interconnect the WE compartment with the membrane sampling compartment. The clcctrolyte flows through the membrane interface sampling compartment via four small capillaries, flowing from the outer edge of the compartment to the centre when it then exits the cell

3.3.1.1 Dual-Cyclone Flow Block

The flow of electrolyte through the dual-cyclone thin-layer flow cell, including the flow regime over the WE and membrane interface is depicted in Fig. 3.7 .

In this design, the electrolyte first enters a capillary leading to the WE compartment, where the electrolyte rises from an inlet located in the centre of the compartment. Electrolyte then flows upwards into the centre of the WE and outwards, in a radial fashion, before exiting the WE compartment via four exit capillaries. These capillaries then transport the electrolyte towards the membrane interface sampling compartment. In contrast to the flow over the WE, the electrolyte enters the sampling compartment from four inlet capillaries located at the perimeter of the volume, and flows towards a central exit capillary that leads to the cell outlet. This flow regime is in principle identical to that described previously in Ref. [2] and furthermore modelled in Ref. [3, 4].

3.3.1.2 Dual-Linear Flow Block

The flow regime through a second flow through module, the dual-linear thin-layer cell is depicted in Fig. 3.8 .

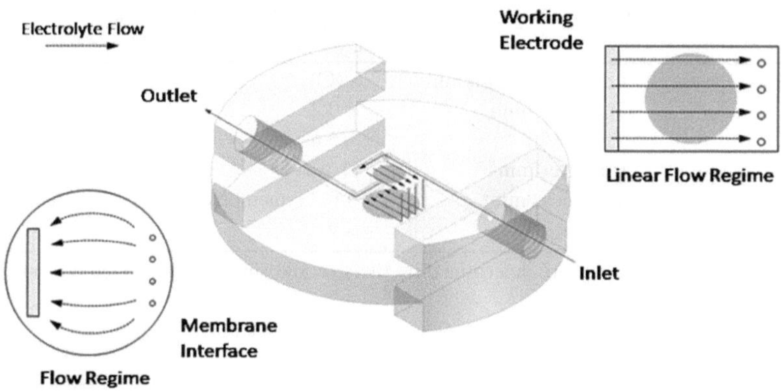

Fig. 3.8 Illustration of the electrolyte flow regime through the DEMS Dual-linear thin-layer electrochemical cell flow through block. The electrolyte first enters a small volume which distributes the flow over the WE surface in a linear fashion. The electrolyte then passes through four capillaries into the membrane interface where it then flows to the cell outlet

In this flow regime, the electrolyte flows through the inlet into an initial 'dead' volume before then flowing into the thin-layer WE compartment in a linear fashion over the electrode surface. The electrolyte then passes through four capillaries, transporting it into the membrane interface thin-layer compartment and passes over the membrane interface before exiting the flow cell. This linear flow cell regime over the WE is not unlike that described previously [5]. The advantage of the linear flow regime is that it is comparatively common in contrast to the dual-cyclone flow regime and can be modelled; essentially, the flow regime over the linear cell is more defined and better understood.

The dimensions and volumes of the DEMS cells are summarised in Table 3.1. The small volumes and narrow connecting capillaries of the cells are a consequence of a number of design considerations. Firstly, by minimising the compartment and capillary volumes, the quantity of solution available for reaction species that arise from the WE electrode to diffuse away and dilute into, is minimised. This, therefore, maximises the concentration of the analyte species entering the membrane sampling compartment which in turn, helps to enhance the flux of diffusing species to the membrane interface surface whilst minimising the delay between the creation of species at the WE and subsequent detection in the QMS. Moreover, the residence time of the species in the electrochemical cell after the electrochemical reaction has ceased is also reduced, helping to improve the correlation between the electrochemistry and mass spectrometry. These narrow connecting capillaries and small cell volumes, however, do have a detrimental effect on the electrochemistry (discussed later in Sect. 4.1.1) in terms of potential control and stability. This necessitates the modification of the standard three-electrode electrochemical setup, employing a stabilising resistor and NID, later described in detail in Sect. 3.6.1.1.

Table 3.1 Details of the specifications of the two DEMS cell flow regimes

		Cyclone	Linear
WE comp.	Length × Width/mm	6 (∅)	7 × 5
	Thickness/mm	~0.2	~0.2
	Volume/μl	~5.6	~7
Sampling comp.	Diameter/mm	11 (∅)	11 (∅)
	Thickness/mm	~0.2 (-membrane)	~0.2 (-membrane)
	Volume/μl	~9	~9

The compartmental volumes are determined by the size and shape of the punctured holes in the cell PTFE gaskets

To summarise, the modular DEMS electrochemical flow cell design presented here allows controlled electrochemistry at the WE of interest and incorporates a PTFE membrane that interfaces the cell electrolyte with the vacuum of a QMS. This is achieved by employing a dual thin-layer flow cell (using either dual- or linear-flow regimes) of modular design that allows easy modification of cell parameters, the exchange of WE substrate or sample, and replacement or alteration of the cell components, whether that is to modify the electrolyte flow regime or membrane type. The performance and characteristics of the dual-cyclone thin-layer electrochemical cell setup are meanwhile presented and evaluated in Sect. 4.1.

3.4 Membrane Interface

All previous DEMS instruments have utilised a microporous PTFE membrane to interface the electrochemical cell with the vacuum system of the QMS (see Chap. 2 Sect. 2.2.2) and this DEMS instrument is indeed no different, simply because microporous PTFE membranes are most suitable for this application. Such materials have a wide range of applications in electronics, pharmaceuticals, laboratories and manufacturing, and are consequently commercially available. The key to designing and constructing the membrane interface is to obtain a suitable commercially available membrane material, define an appropriate geometric area and decide how to form and seal the membrane between the electrochemical cell and vacuum flange .

The first and foremost requirement of the microporous PTFE membrane is that it must not be permeable to aqueous electrolyte, which can in principle be ensured by choosing a membrane pore radius <0.8 μm [6]. Nevertheless, an appreciable amount of aqueous electrolyte still evaporates through the microporous membrane owing to a significant pressure gradient across the membrane interface, as was highlighted previously in Chap. 2 Sect. 2.2.3. The pore size itself, however, does not tell us too much about the total flux of gas that will pass through the DEMS membrane interface when exposed to the vacuum, which is actually the crucial parameter required to design the DEMS membrane interface and differentially pumped vacuum system. Unfortunately, the total gas flux through a microporous

Table 3.2 Details of the microporous PTFE membrane interface material candidates obtained for this DEMS instrument construction

Manufacturer	Ref no.	Pore size/μm	Thickness/μm	Water entry pressure/kPa
Gore	MMT-316	0.1	127 ± 9	4.8
	MMT-335	0.03	124 ± 11	16.1
Donaldson	#1302	0.2	89	620*
	#6860	0.4	152	517*
	#6865	1.5	152	117*
	#1301	0.1	89	895*
	#1314	1.0	89	140*
Whatman	–	0.2	–	–

Note water entry pressures Marked with * Mullen-Hydrostatic values

membrane separating aqueous electrolyte and a vacuum is not a specification of the commercial membranes, although it will be determined by the membrane properties i.e. pore size, pore density, thickness, and geometrical area, not to mention the temperature of the aqueous electrolyte and pressure gradient across the membrane. The only conceivable approach toward obtaining an appropriate membrane material, therefore, is to source a number of commercial microporous PTFE membrane samples and test their suitability.

A number of potentially suitable commercial microporous PTFE membrane samples were obtained from three manufacturers, whose specifications (namely pore size) were chosen based on the details provided in previous DEMS literature, which were summarised previously in Table 2.1 Sect. 2.2.2. The manufacturer specifications of the membrane samples obtained for this DEMS instrument are presented in Table 3.2.

The material obtained from both *Gore* and *Donaldson Membrane Solutions* were conveniently available in A4 sheets allowing the desired shape and size of the membrane to be cut out (using a hole-punch, for example). There are then two possible approaches to testing the suitability of the membrane candidates. The first approach, which was taken in this project, is essentially trial and error, whereby an electrochemical cell and vacuum system is designed and constructed, and a range of membrane materials are obtained and tested. Either the electrochemical cell or vacuum system (or even both) is then modified to achieve a balanced system with appropriate vacuum pressures and sensitivity. Another approach would have involved constructing a membrane test setup using a small vacuum chamber, pressure gauge and roughening pump, using which the gas flux through the membrane is determined in an experiment similar to that described later in Sect. 4.2. Once the gas fluxes have been established, it is then possible to choose upon a suitable membrane and/or geometric area, and then design the vacuum system (or vice versa) based on the expected total gas flux. In hindsight this latter approach would have perhaps been somewhat wiser, but came rather as an afterthought.

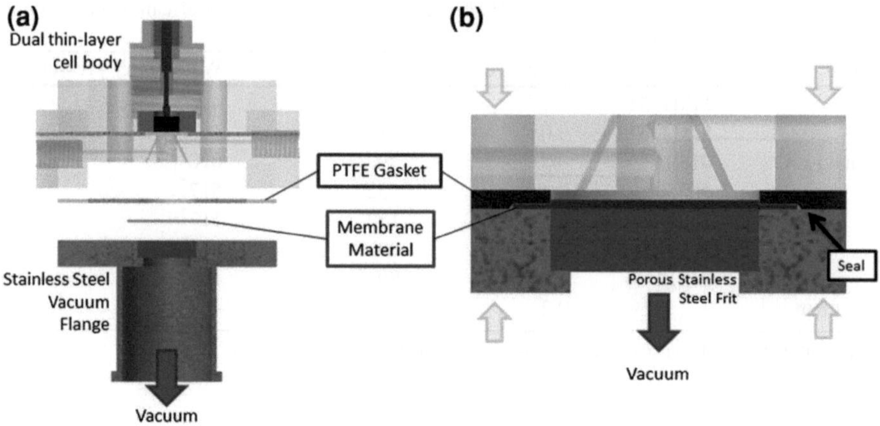

Fig. 3.9 An illustration of the membrane interface. The microporous membrane material is mechanically supported by a stainless steel frit, and sealed between the cell flow through block and vacuum flange via a PTFE gasket. When compressed the PTFE sheet forms a gas tight seal around the edges of the membrane material between the PTFE gasket and vacuum flange connection

The incorporation of the membrane material in the DEMS electrochemical flow cell is illustrated in the electrochemical cell cross-section of Fig. 3.9. The membrane is located between the PTFE membrane compartment gasket and the vacuum flange, and is positioned so that it covers the porous stainless steel frit entrance to the vacuum chamber. When the electrochemical cell components are combined, fixed and PTFE gaskets compressed using the four M4 bolts (which fix into the vacuum flange) the membrane sampling compartment volume is formed, as depicted in Fig. 3.9.

In practice there are a few crucial points to consider when dealing with the membrane interface: firstly, it is absolutely necessary that the membrane completely covers the porous metal frit to prevent aqueous electrolyte from entering the vacuum chamber. Secondly, it is equally important that the PTFE gasket is compressed against the flange so that it forms a seal around the membranes circumference. The compression of the membrane material against the flange alone is not sufficient because the membrane backing material is still highly gas porous. Any additional gas leaking through the membrane will add to the total gas flux entering the vacuum system resulting in higher pressure and larger QMS mass ion current background signals. The third necessity, is that the aqueous cell electrolyte completely covers the membrane surface before the DEMS vacuum system is exposed to the membrane so that the vacuum is formed i.e. without the aqueous electrolyte (forming a seal) air will simply pass through the membrane with little or no resistance, which could potentially destroy the vacuum equipment.

Ideally, it is desirable to maximise the geometric area of the membrane to volume of electrolyte ratio in order to attain as high a sensitivity by collecting as much of the electrochemical reaction species as possible. However, the larger the geometrical area of the membrane is, the greater the gas flux will be and therefore the more vacuum system pumps will be required.

3.5 Vacuum System Design

The purpose of the DEMS vacuum system is to achieve the high vacuum conditions required for the reliable operation of the QMS. In terms of the DEMS instrument, the vacuum system must be capable of sufficiently pumping away the total flux of gas that passes through the membrane interface at such a rate that the operating pressure of the QMS is maintained. The key to constructing a vacuum system of appropriate design, therefore, was to balance the constant flux of gas entering the vacuum chamber via the membrane interface, against the overall pumping speed of the vacuum system. This was achieved by designing and developing the three-stage differentially pumped vacuum system illustrated in Fig. 3.10, with each of the three chambers highlighted in green, blue and yellow.

The three-stage differentially pumped DEMS instrument vacuum system possesses three connected vacuum chambers that each operate at different vacuum pressures. These are evacuated by individual turbo molecular pumps (two 70 $l\ s^{-1}$ and a single 300 $l\ s^{-1}$) that are backed by two rotary vane pumps. The electrochemical cell including membrane interface, is attached via a KF flange to a leak valve connection to the vacuum system. An exploded diagram of the 3-stage differentially pumped vacuum system created in this work is depicted in Fig. 3.11.

The leak valve connecting the electrochemical cells membrane interface is attached to a 1st pumping stage, evacuated by a relatively small 70 $l\ s^{-1}$ turbo molecular pump, backed by a single rotary vane pump. The 1st pumping stage is then coupled to a 2nd pumping stage whose cavity contains the ion source and front section of the quadrupole rods of the QMS, which is evacuated using a larger 300 $l\ s^{-1}$ turbo molecular pump. The relative pressure difference between these two chambers is determined by the conductance of gas passing from the 1st into the 2nd pumping stage, which can be controlled using an aperture of a certain dimension. An additional 3rd pumping stage (highlighted in yellow) is located at the 90° axis between the quadrupole and SEM of the QMS, and is separated from the 2nd chamber using a flange that partially encloses the cross-section of the outer quadrupole casing. This 3rd pumping stage lowers the pressure at the SEM relative to the 2nd pumping stage using another 70 $l\ s^{-1}$ turbo molecular pump, and allows a slightly higher gas pressure at the ion source relative to the SEM.

The key features that differentiate this vacuum system from previous DEMS instruments, is the use of three differentially pumped vacuum chambers, and the tubular aperture which is orientated at a 90 ° angle with respect to the quadrupole axis and ion source. Further description and details of these features, namely the

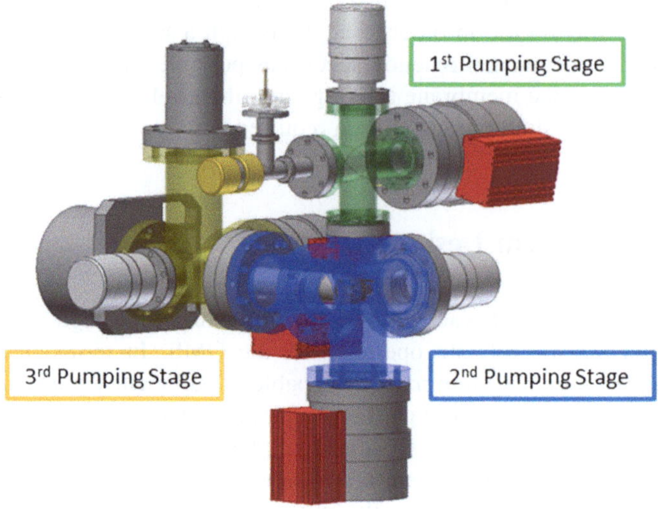

Fig. 3.10 An illustration highlighting the chambers of the 3-stage differentially pumped vacuum system. The 1st pumping stage operates at a rather high vacuum pressure and is connected via a tubular aperture to the 2nd pumping stage containing the ion source. The 3rd pumping stage meanwhile is located in the chamber connecting the quadrupole and SEM of the QMS

three-pumping stages and aperture will be presented in the following two sections, respectively.

3.5.1 Three-Stage Differential Pumping

The pressure requirements of the vacuum system design are essentially defined by the upper operating pressure limits of the QMS ionisation, mass ion selection and detection processes. The reliable upper operating vacuum pressure limit provided by the manufacturer varies slightly for each component of the QMS, and is summarised in Table 3.3.

The absolute upper operating pressure limit is dependent upon the QMS detection mode, either using the faraday cup, or secondary electron multiplier (SEM). If the whole QMS is positioned in a single vacuum chamber, then the upper operational pressure limit is determined by the QMS mass ion detection mode. For measurements where the detected ions are measured using the faraday cup (during which time the SEM is turned off), the upper operating pressure is 1.0×10^{-4} mbar, whereas for SEM operation, the upper operating pressure limit is much lower at 5.0×10^{-6} mbar. These pressures are easily achievable in a

Table 3.3 Details of the QMS component operating pressure limits [20]

QMS component	Operating pressure/mbar
Cross beam ion-source	$<1.0 \times 10^{-4}$
Quadrupole	$<1.0 \times 10^{-4}$
Faraday cup	$<1.0 \times 10^{-4}$
Secondary electron multiplier (SEM)	$<5.0 \times 10^{-6}$

When operating the QMS with the SEM the vacuum pressure must be nearly two orders of magnitude lower than with the faraday cup

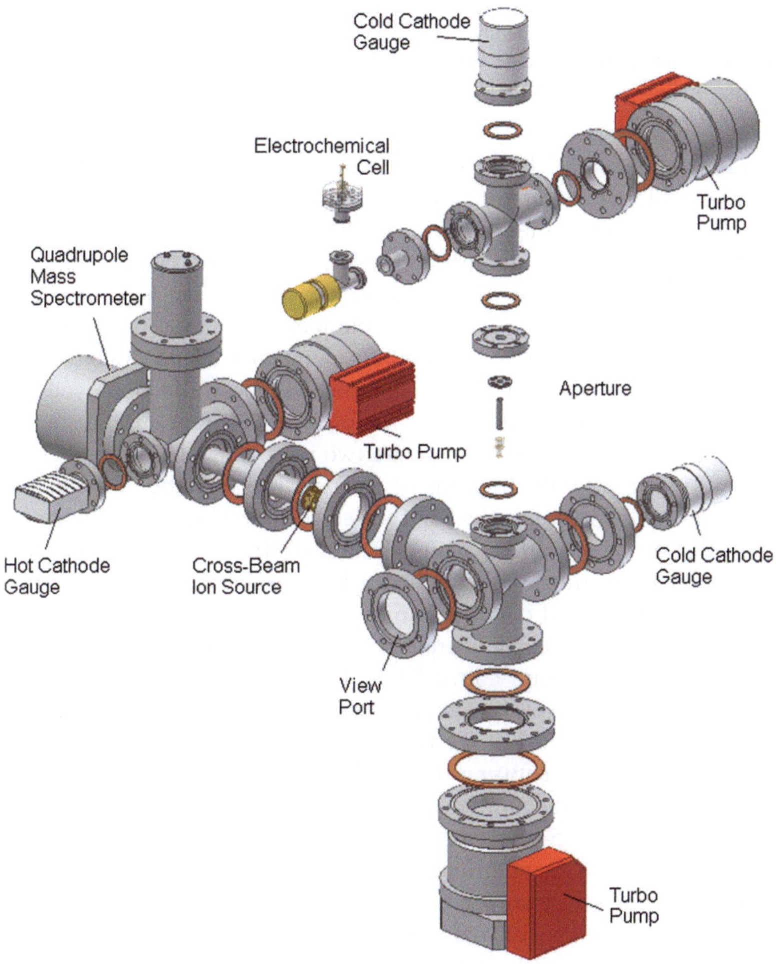

Fig. 3.11 Exploded diagram of the DEMS instrument vacuum system and QMS. The three pumping stages are connected via two apertures which control the flux of gas between each chamber to achieve the necessary differential pumping. Each chamber is evacuated by a turbo molecular pump whereby each pressure is measured by a vacuum gauge

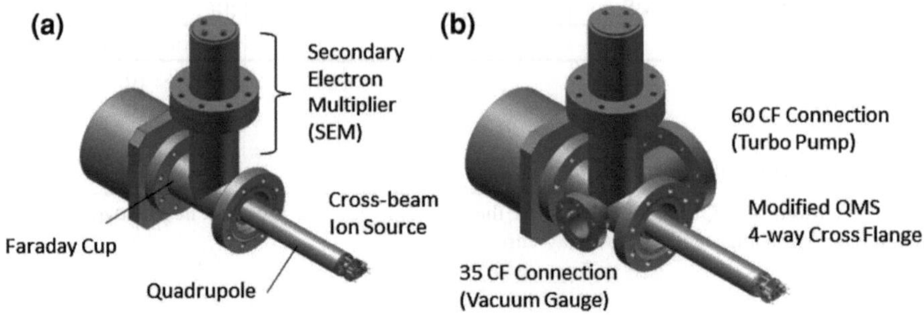

Fig. 3.12 Illustration of the original 3-way flange QMS vacuum chamber housing (**a**) and modified 5-way QMS vacuum chamber housing (**b**). The original QMA 422 vacuum chamber was adapted to incorporate a flange connection for the turbo pump and pressure measurement gauge allowing the 3-stage differentially pumped vacuum system design

closed vacuum chamber, however, in DEMS we require a constant gas flux through the membrane to continuously enter the chamber, but be pumped away at such a rate that the operating pressures of the QMS are maintained.

In order to achieve these pressures using a single vacuum chamber there are two possibilities. Either, the gas flux through the membrane is controlled to an acceptable value by adjusting the size of the membrane inlet (possible with capillary inlets), or we employ a differentially pumped vacuum system. Two-stage differentially pumped vacuum systems have been successfully employed in DEMS previously [3, 7–9], which poses the question why is it that a three-stage system has been constructed here. Well, there are two reasons related to the performance and reliability of the QMS. In terms of performance or rather, sensitivity of the QMS, it is desirable to employ the SEM which considerably amplifies the mass ion current, whilst at the same time maintaining a relatively high pressure at the ion source in order to maximise the number of ions produced. In terms of reliability, however, the SEM should only operate at vacuum pressures $<5.0 \times 10^{-6}$ mbar, whereas the CB-ion source can operate at pressures up to 1.0×10^{-4} mbar. If the SEM is operated at pressure above 5.0×10^{-6} mbar then the performance and lifetime of the SEM is severely compromised [10]. A pressure difference between the CB-ion source and SEM of the QMS is therefore desirable.

A pressure difference between the CB-ion source and SEM was achieved by modifying the original vacuum housing of the QMS. This modification from the original chamber to the 5-way cross chamber is depicted in Fig. 3.12.

The addition of two vacuum flanges to the QMS housing allows an additional turbo-pump and pressure gauge to be placed between the SEM and ion source. An additional flange is then positioned over the quadrupole to limit the conductivity of the vacuum tube in order to achieve a larger pressure difference between the CB-ion source (2nd chamber) and SEM (3rd chamber). The resulting vacuum pressures, calculated tube conductances and effective pumping speeds are provided in the schematic in Fig. 3.13.

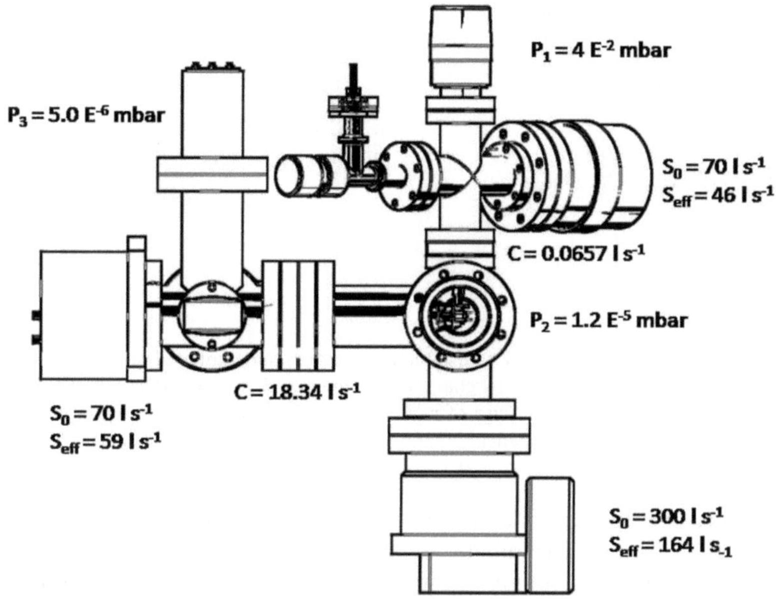

Fig. 3.13 A sketch of the DEMS vacuum chamber giving the respective pumping speeds (*S*), conductances (*C*) and operating pressure (*P*) of the 3-stage differentially pumped vacuum system when the DEMS instrument is in operation. The effective vacuum pumping speeds and gas conductance of the apertures are determined by the internal dimensions of the components

The turbo molecular pumps located in the 2nd and 3rd pumping stages are backed using a single rotary vane pump. The vacuum pressures are monitored by two cold cathode gauges in the 1st and 2nd chamber, and a hot cathode gauge located in the 3rd chamber. Once the DEMS electrochemical cell is attached to the vacuum system, and the valve exposing the membrane interface is fully opened, a constant gas flux into the vacuum chamber results. This gives typical steady operating vacuum pressures of $\sim 10^{-2}$, $\sim 2 \times 10^{-5}$ and $\sim 5 \times 10^{-6}$ mbar in the 1st, 2nd and 3rd chambers, respectively.

3.5.2 Tubular Aperture

The tubular aperture of the DEMS instrument in this design serves two essential functions. Its primary function is to limit the conductance of gas between the 1st and 2nd differentially pumped vacuum chambers so that a considerable lower pressure in the 2nd chamber is achieved. Its second function is to guide the flux of gas originating in the first chamber into the CB-ion source of the QMS. This was achieved using the adjustable tubular aperture design depicted in Fig. 3.14.

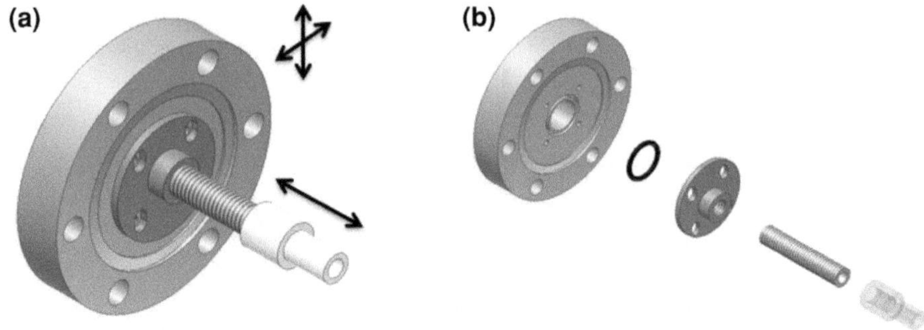

Fig. 3.14 Illustration of the DEMS adjustable aperture (**a**) and exploded image (**b**). The design allows the conductance to be varied by changing the internal diameter and length of the tube. The adjustable nature of the design and use of a Macor® tip allows the aperture outlet to be position closely to the CB- ion source entrance

The adjustable aperture is mounted onto NW35 CF vacuum flange, and sealed using a Viton O-ring. The tubular section of the aperture is screwed into the mounting plate and is therefore adjustable in the vertical direction by screwing or unscrewing the tubing. The mounting plate itself also possesses oval holes allowing some adjustment in the horizontal direction. The exact dimensions of the aperture, most significantly the internal diameter and length of the tube determines its gas conductance. The dimensions of the tubular aperture can therefore be tuned to give the necessary differential pressure between the 1st and 2nd pumping chambers. The aperture used here is 5 mm internal Ø and 8 mm in length, allowing a gas flux of $\sim 0.07\ \text{l s}^{-1}$. Finally, the tubular aperture is capped with the glass ceramic material, Macor®, to prevent discharging from the CB-ion source to the conducting aluminium part of the aperture, so that the aperture may be positioned in close proximity.

By positioning the aperture close to the CB-ion source, and therefore directing the molecular beam directly into the CB-ion source, the partial pressure of gas leaking into the 2nd chamber through the ion source is increased with respect to the surrounding vacuum chamber. The adjustability of the aperture is important therefore, because the positioning of the tubular aperture over the cross-beam ion source can have a significant effect on the performance of the QMS. There are two possible entrances into the CB-ion source as shown in Fig. 3.15.

The molecular beam exiting the aperture may be directed either along (indicated with the red) or perpendicular (indicated in blue) to the quadrupole axis in order to enter the CB-ion source ionisation area (Wehnelt Cage). In contrast to the smaller front entrance (indicated in red) the perpendicular entrance (either above or below the Wehnelt cage) is much larger, and the aperture can more easily be positioned correctly. The protruding molecular beam will therefore more easily pass through the CB-ion source rather than hitting the outer face of the Wehnelt cage and bouncing off into the surrounding vacuum chamber. In this DEMS instrument therefore, the aperture and protruding molecular beam was orientated

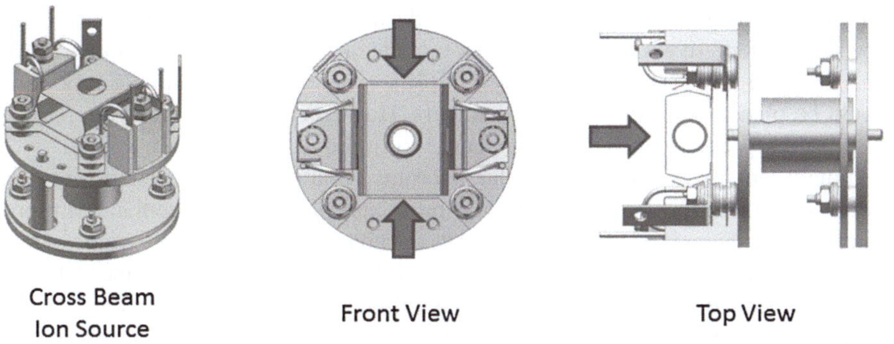

Fig. 3.15 Illustration of the QMS cross-beam ion source highlighting possible molecular beam inlet orientations. The Wehnelt chamber of the ion source possesses two possible entrances which may be either in-line or perpendicular to the axis of the quadrupole. The perpendicular entrance is rather more open and was the chosen orientation of the aperture used in this DEMS instrument

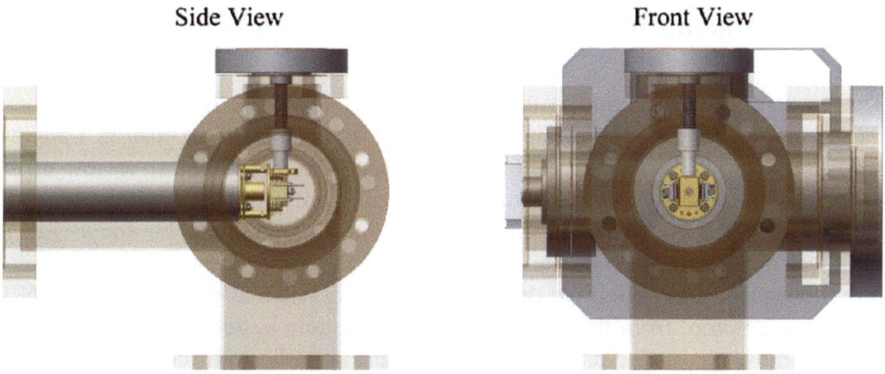

Fig. 3.16 Technical drawing indicating the position and orientation DEMS vacuum aperture with respect to the cross-beam ion source. The aperture outlet is positioned directly over the ion source directing a large proportion of the gas entering the 2nd vacuum chamber through the ion source. This effectively raises the pressure of the ion source enhancing the number of ionised species, increasing the sensitivity of the QMS. Aperture orientation wrt. CB-ion source

perpendicularly to the quadrupole axis. This is as depicted in Fig. 3.16 where the exit of the tubular aperture (Macor® tip) into the 2nd pumping stage is positioned directly over the entrance to the CB ion source.

A tubular aperture is not necessarily the only possible solution because a simple hole would also suffice in order to control the conductance of gas between chambers. The tubular aperture, however, not only possess the advantage that the protruding molecular beam can be position directly over the CB ion source but also that is more easily fabricated compared to the simple hole i.e. the exact conductance is sensitive to the precise dimensions of a hole aperture with little room for error, whereas a tube of a certain diameter may be simple adjusted in length. In cases where it would be perhaps desirable to tune the conductance of the aperture, alternative design solutions include an inexpensive variably control aperture designed in [11] and realised in [12], or alternatively a simple leak valve could be placed between vacuum chambers. However, these solutions do not help to raise the pressure at the CB-ion source to improve the QMS sensitivity.

Finally, in cases where a tubular aperture separates vacuum chambers with differential operating pressures of ca. 1 mbar and possess a molecular beam protruding from the exit of the tube as the gas expands into the lower pressure vacuum chamber, it is possible that the sensitivity of the QMS could be enhanced further by the addition of a chopper and lock-in amplifier to reduce measurement 1/f noise [13]. This more complicated system, however, was not employed in this experimental setup. Another alternative would have been to employ a closed ion source design which are commercially available [14] and can increase the sensitivity of the QMS quite considerably.

3.6 Instrumentation, Control and Data Acquisition

The purpose of this section is to present the details of the hardware instrumentation, control and data acquisition (DAQ) devices, as well as custom Labview software setup of the DEMS instrument. Its aim is to provide an overview of the most important aspects of the hardware and software setup, and when necessary guide the reader to where further, relevant information may be found i.e. in technical notes and manuals etc. The hardware instrumentation can be separated into those that are required for the electrochemistry, such as the potentiostat, NID and syringe pump, from those required by the vacuum system and QMS, such as turbo pumps, pressure control units and signal amplifiers. In common to many pieces of the DEMS instrument hardware, however, is the use of DAQ cards and software to both control instrumentation and record measurement data.

Whilst historically, analogue function generators and xy-plotters were essential in electrochemical experimentation, such devices ought to have no place in a state-of-the-art electrochemistry laboratory. Data acquisition hardware and graphical software programming tools such as Labview offer electrochemists the possibility to meticulously control potentiostats amongst other experimental parameters programmatically, with precise data acquisition, real time data plotting and recording. Although commercial electrochemical instrumentation software is available, their functionality or customisability cannot be compared to a well-designed Labview based electrochemistry software. Furthermore, custom data treatment software that performs common treatment tasks can also allow rather efficient post-measurement data analysis. Hence, a customised Labview software was the approach taken in our electrochemistry laboratory, further details of which may also be found in Ref. [15]. In short, the role of the traditional analogue function generator is replaced by a software generated digital waveform that is output via a digital to analogue converter (DAC) device. Computerised rather than manual control allows experimental measurement parameters and techniques to be performed meticulously, with defined measurement sequences and protocols. Whereas previously, analogue signals were recorded and displayed using an X–Y recorder, plotter or oscilloscope, data is almost exclusively today transferred to computer using an analogue-to-digital converter (ADC) device. The data can then be easily displayed and recorded in an appropriate format that is tuned to the measurement type. This is particularly useful in electrochemistry were a single CV can measured relatively quickly, and consequently a large quantity of data can be accumulated in a relatively short period of time, making it equally important to be able to analyse measurement data efficiently. The advantages of a custom Labview programmed instrument control and data acquisition led to the extension of the approach taken with the RDE and MWE methodologies, to the DEMS instrument. Overall, the instrument requires two measurement setups, one for DEMS experimentation and another for QMS calibration.

3.6.1 DEMS Measurement Setup

An overview of the communication between the various instrumentation, control and data acquisition hardware and software required to operate the DEMS instrument acquire measurement data during experimentation is presented in Fig. 3.17.

The instrumental setup relies on two sets of software which control or record data from three instruments (the syringe pump, potentiostat and QMS control unit) using either the two DAQ cards, or RS232 data communication. The first piece of software is the commercial Quadstar 32-bit program which is used simply for the operation of the QMS. The second piece includes the custom developed Labview software programs: Macro and CV which control, measure, display and save the DEMS electrochemical data. These applications were originally created for RDE and MWE methodologies; however, as part of this thesis the software was developed further and extended to incorporate the additional functionality required to acquire data from the QMS. Both the electrochemical and QMS experimental data is consequently recorded, treated, synchronised, displayed and saved in real-time. Finally, the Macro software also incorporates programmatic control over the electrochemical flow cell syringe pump for the control of the electrolyte flow rate. Overall, this custom Labview program approach considerably increases the functionality and usability of the DEMS instrument.

3.6.1.1 Electrochemistry Hardware

The DEMS electrochemical flow cell setup relies on the control of three pieces of hardware: the syringe pump to control the electrolyte flow rate through the electrochemical cell, the potentiostat to control the WE potential in electrochemical experimentation, and finally, the NID to achieve *iR*-compensation.

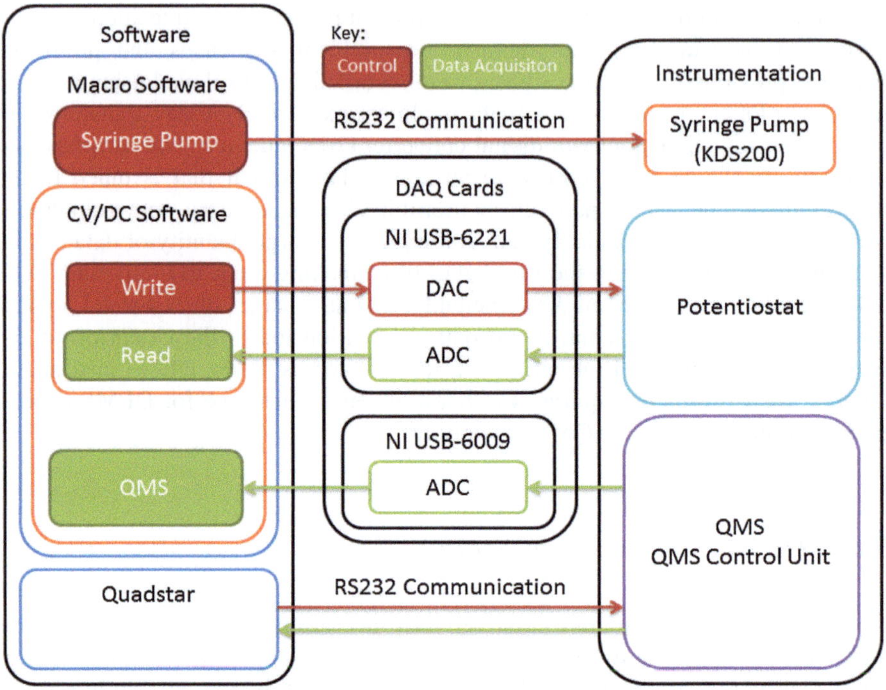

Fig. 3.17 Overview of instrumentation, control and data acquisition hardware and software employed in this DEMS instrument. The custom electrochemistry software is used to control the potentiostat and syringe pump for the electrochemical cell, and acquire data from the potentiostat and QMS. The commercial 32-bit software meanwhile is used to exclusively operate the QMS

Syringe Pump

The electrochemical flow cell setup employed in this DEMS instrument is of a flow cell nature and therefore requires the continuous passage of electrolyte through the electrochemical cell at a precisely controlled rate. In this setup, this is achieved by placing a syringe pump on the flow cell outlet and operating it in withdrawal mode which limits the flow of electrolyte through the cell to a certain rate. The syringe pump employed is a KDS 200 infusion withdrawal pump using either a 25 or 50 ml syringe with typical withdrawal rates of between 60 and 600 μl min^{-1}. Refer to the syringe pump manual for further details on its operation [16]. Although manual control of the syringe pump during DEMS experimentation is of course possible, in certain experiments it can be desirable to have computer control over the syringe pump which can be achieved using RS232 communication, and Labview VIs which were incorporated into the custom Macro program. The flow control software is later described in Sect. 3.6.3.

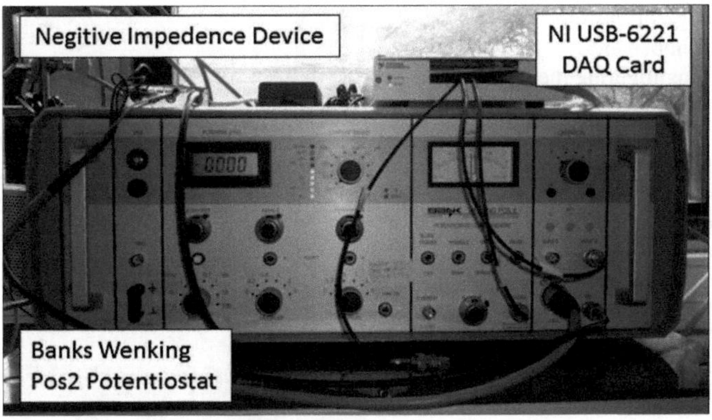

Fig. 3.18 Image of DEMS electrochemical potentiostat hardware setup. The custom NID device is used to apply *iR*-compensation, whilst the DAQ card is used to both control the potentiostat via the custom Labview electrochemistry software and measure the WE potential from the potentiostat and electrode current via the NID

Potentiostat

The potentiostat employed in this DEMS instrument is a Banks Wenking POS 2 potentiostat which may be controlled either manually using the internal scan generator, an external function generator, or finally via computer control using the SCSI interface. In this setup depicted in Fig. 3.18, the potentiostat is controlled via its potential control analogue input by the DAQ which effectively functions as an external function generator that is controlled by the CV and Macro software.

The DAQ simultaneously applies the desired WE potential and measures the corresponding measurement potential and current. The current range meanwhile is controlled using the DAQ card digital outputs via the SCSI interface on the rear of the potentiostat. The instrument operation mode i.e. whether it is set to off, open circuit potential, potentiostatic or galvanostatic, however, must be performed manually. In principle, the potentiostat and control system is identical to the RDE setups used in the rest of our electrochemistry laboratory, with the exception that a NID device is incorporated into the electrochemical cell setup. The NID device is required for two reasons: firstly, the positive feedback *iR*-compensation scheme of this particular potentiostat does not function properly, and secondly, the NID circuit has some advantages over positive feedback schemes to compensate highly resistive electrochemical cells.

Negative Impedance Device

The negative impedance device (NID) is an operational amplifier based circuit that is positioned between the WE of the electrochemical cell and potentiostat WE input connection. An overview of the in-house built NID and its connections is depicted in Fig. 3.19.

The circuit functions as a 'negative resistor', and adjusts the potential applied to the actual WE relative to the potentiostat output, according to the current that

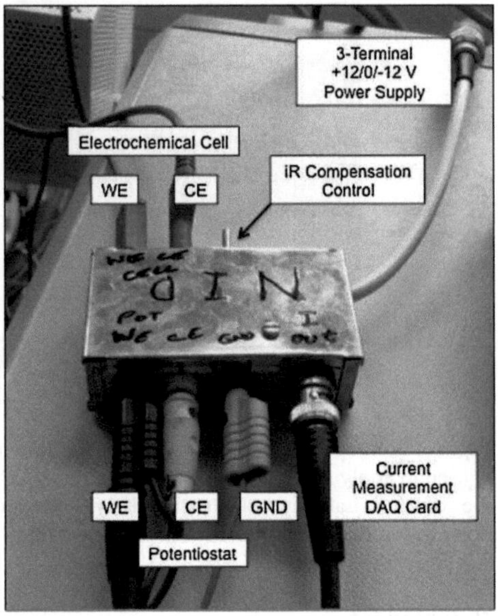

Fig. 3.19 Image of custom built negative impedance device (NID) highlighting the potentiostat and cell connections. The NID is located between the potentiostat and the WE electrode, allowing adjustable *iR*-compensation of the solution (or in this case stabilising resistance). In this setup the DAQ measures the current response via the NID rather than the potentiostat

flows through the WE (and therefore NID) and the choice of circuit resistances. A general schematic of the NID circuitry is given in Fig. 3.20. The potential difference across the NID, V_c, is described by the following equation:

$$V_c = -iR_m\frac{R_1}{R_2}$$

In a circuit where R_1 and R_2 are of equal resistances, V_c is purely determined by the resistance of R_m and the current flowing through the WE. By measuring the ohmic resistance between the Luggin-capillary and WE, and then applying an equivalent R_m resistor, complete compensation of the solution resistance, R_s, may be achieved.

This particular NID circuit design, however, is only suitable in circumstances where R_s is both known precisely, and is constant during the course of an entire experiment (or set of experiments) which is not often the case. Furthermore, when using the NID over compensation of the R_s may be possible under certain circumstances, in contrast to the positive feedback circuitry employed in analogue potentiostats, where complete and over compensation results in potentiostat instability and oscillation [17, 18]. The usefulness of the NID device circuit design shown in Fig. 3.20 is therefore limited but would be significantly enhanced if the amount of R_s it compensates could be more easily adjusted.

A modified NID circuit given in Fig. 3.21 was therefore developed (which is contained in the housing shown in Fig. 3.19) for the DEMS electrochemical cell setup and is a modification of the NID designs previously reported [19]. Note that in this case the solution resistance is replaced with an artificial stabilising

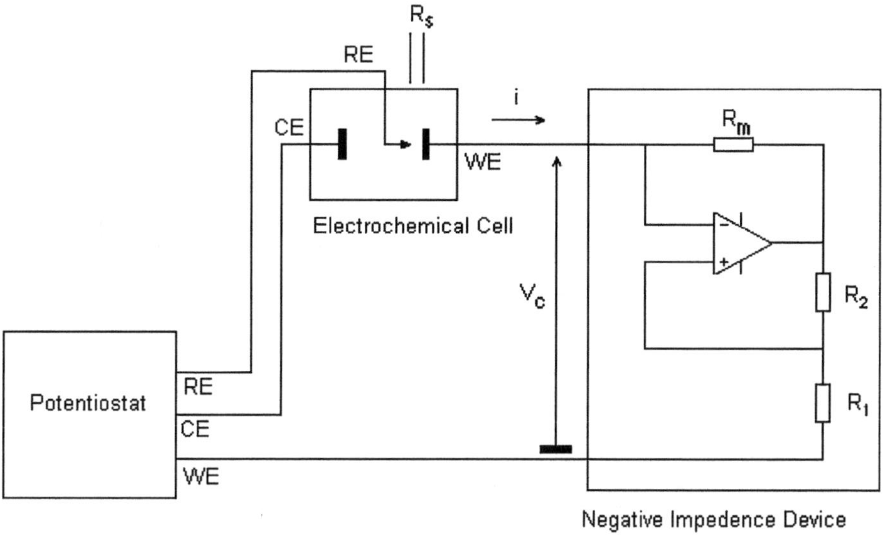

Fig. 3.20 A schematic of the original NID circuit design and electrochemical cell setup. The circuit functions as a negative resistor and can be used to compensate the solution resistance in an electrochemical cell without interference on the potentiostat circuit. The amount of *iR*-compensation is typically determined by the size of the resistor, R_m

resistance, R_{st}, which is necessary to stabilise the DEMS electrochemical flow cell and is later discussed in more detail in Chap. 4 Sect. 4.1.1.

In this circuit, the extent of R_{st} compensation can be adjusted using a variable resistor, R_1, in combination with a fixed, R_2 and R_m resistors. An R_m value of 1.0 kΩ was chosen in order to give a 1.0 V across the resistor when a current of 1.0 mA flows through the WE which can easily be measured using standard DAQ cards. A combination of R_1 and R_2 values of 10 and 50 kΩ resistances, respectively, offers a flexible range of R_s compensation from 0 to 200 Ω using this NID circuit. Another feature is the 0.01 μF filter capacitor, C_1, which allows AC frequencies >10 kHz to bypass R_m so that measurement noise is also reduced.

The introduction of the stabilising resistor, R_{st}, also has the additional advantage that the actual WE potential can be measured using a DAQ channel that directly measures the potential between the WE and RE, indicated as CH1 in Fig. 3.21. The actual potential and current relationship is therefore measured directly, rather than applying *iR*-correction in post measurement data treatment. The post data treatment *iR*-correction of the potential for large uncompensated resistances in a CV for instance, will introduce errors into the measured current, particularly for electrode processes dependent upon scan rate. The direct measurement of the WE potential using the DAQ card in combination with an NID circuit could therefore be a useful compromise in electrochemical experimentation where it is necessary to work with poorly conducting electrolytes. In such circumstances the true current and potential relationship is measured rather than performing post measurement data correction,

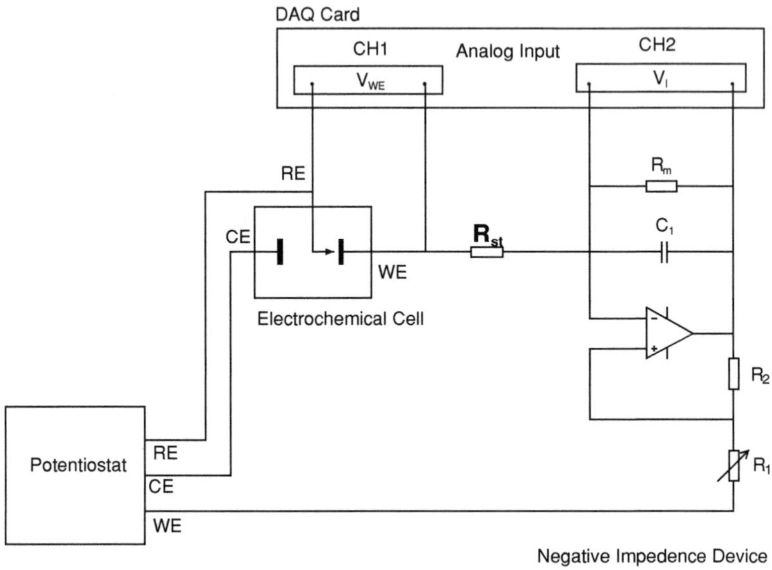

Fig. 3.21 A schematic of an enhanced NID circuit design allowing adjustable *iR*-compensation developed for the DEMS instrument as part of this thesis. The circuit allows the amount of *iR*-compensation of the solution (or stabilising) resistance to be adjusted by the variable resistor (potentiometer) R_1 whilst enabling the WE current to be measured across a fixed R_m resistor using a DAQ card

which for large resistance can be significantly erroneous. The device also allows adjustable *iR*-compensation and measurement of the WE current, without directly interfering with the potentiostat control circuit. It should be noted, however, that the NID can only be used with the potentiostat functioning in potentiostatic mode i.e. *iR* compensation of the WE potential is not possible in galvanostatic operating modes, although for galvanostatic measurements the current is generally not varied but static and thus the post measurement correction of uncompensated solution resistance is more acceptable.

Electrochemical Setup DAQ Card

The potentiostat is controlled from the custom Labview software via a National Instruments USB-6221 DAQ device. This is identical to the DAQ devices used in the RDE setups in the electrochemistry lab, although in this setup the current is measured via the NID rather than the potentiostat current output. The DAQ card possesses two analogue outputs, one of which is used to control the WE potential via the potentiostat (or rotation in the RDE setups), and 16 analogue inputs, two of which are used to measure the applied WE potential and current. The digital outputs meanwhile are used to programmatically control the potentiostat current measurement range. Although this DAQ card could also have been employed to acquire analogue data from the QMS, a separate inexpensive DAQ card was

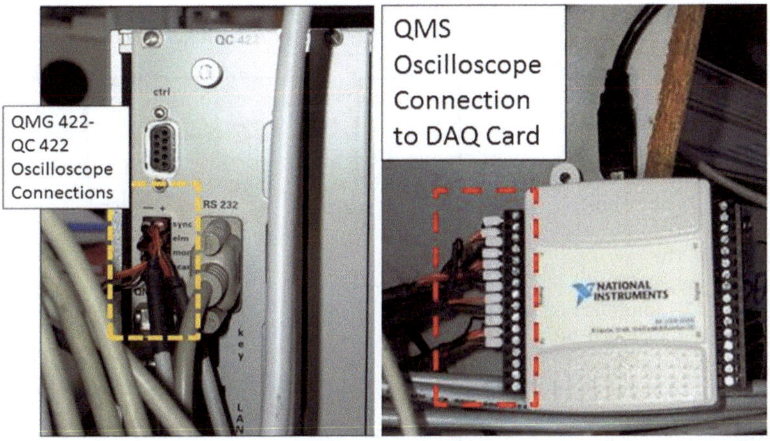

Fig. 3.22 Images depicting the QC 422 oscilloscope connections (*left*) and NI DAQ card used for QMS data acquisition (*right*). The DAQ card is used to acquire the QC 422 oscilloscope voltage output which is then treated by the custom Labview software to acquire the measurement data

instead used to simplify the software programming and maintain the electrochemistry software architecture, described later in Sect. 3.6.3. For more details of the electrochemistry DAQ setup refer to Ref. [15].

3.6.1.2 Quadrupole Mass Spectrometer Hardware

The operation of the QMS (QMG 422/QMA 400) mass spectrometer requires a number hardware including the QMH 400 high frequency generator, EP 422 electrometer preamplifier and the QMA 422 control unit that make up the whole QMS. The average user would perhaps not need to be aware of most aspects of this hardware, which is simply controlled using the Quadstar 32-bit software. However, the DEMS user should be aware of the method in which QMS data is acquired from the instrument and the importance of choosing the correct amplifier range for the mass ion current .

***QMS* (*QMG* 422)**

Data from the QMS may be obtained in real-time using either DDE communication between Labview and Quadstar, or via the oscilloscope connections located on the back of the QC 422 module, as indicated in Fig. 3.22(left) (Note that DDE communication is software rather than analogue based and is discussed in more detail in Sect. 3.6.2.2). The DAQ hardware employed to acquire the QMS analogue signal is National Instruments USB–6008 card with 8 analogue inputs, as depicted in Fig. 3.22(right).

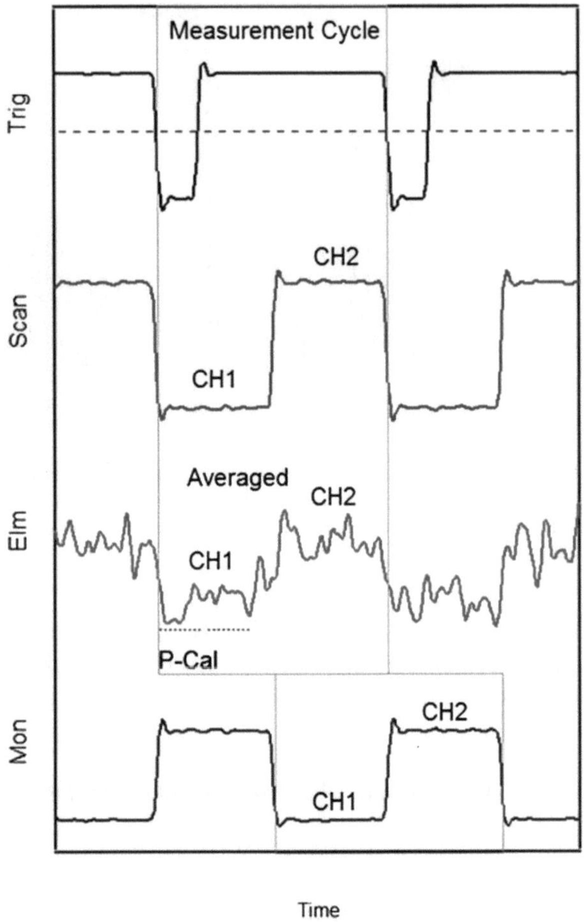

Fig. 3.23 Example QMS analogue measurement signal outputs (intended for an oscilloscope) of the QC 422 for the measurement of two QMS channels. The *Trig* output defines the start of a measurement cycle, whilst *scan* corresponds to the selected mass. The raw electrometer output (*Elm*—from the EP 422 amplifier) is then treated and averaged by the QC 422 unit which outputs a corresponding fixed value (*Mon*). Note that the *Mon* output is delayed by one measurement channel

The analogue output signals of the QC 422 unit are labelled trig (trigger), elm (elementary), scan (scan) and mon (monitor). The exact nature and composition of these analogue output signals are dependent on the measurement mode of the QMS i.e. whether in mass scan or mass ion detection modes. Because the mass ion detection mode of the QMS is employed in DEMS, only the analogue outputs in this mode will be explained here, however, a more comprehensive overview is given in Ref. [20]. An example analogue signal output of the QC 422 unit during the mass ion detection (MID) of two different mass ion ratios is depicted in Fig. 3.23 .

Each fall in the trigger analogue output voltage (from 5 to 0 V) indicates the onset of a new QMS measurement cycle. The selected mass ion current channels (defined in the Quadstar 32-bit software) are then cycled through, as indicated by a corresponding change in the scan output voltage which is proportional to the selected mass ion charge ratio of the QMS. The change in measurement channel

Table 3.4 EC 422 Electrometer amplifier range (AO = Analogue Output)

AO range	Measurement range/A	Sensitivity/A V^{-1}	Rise time/Ms
10^{-5}, 10^{-6}	10^{-5}	10^{-6}	50
10^{-7}, 10^{-8}	10^{-7}	10^{-8}	90
10^{-9}, 10^{-10}	10^{-9}	10^{-10}	1.9
10^{-11}, 10^{-12}	10^{-11}	10^{-12}	2.6

The EP 422 device only possesses four physical amplification ranges, despite that the Quadstar 32-bit software allows one of eight ranges to be selected

mass corresponds to a change in the raw amplifier output signal (*elm*). In the QC 422 this signal is averaged and output in the monitor analogue output delayed by a single channel. The monitor output voltage signal is determined by both the channel mass ion current, and the analogue output setting defined in the Quadstar 32-bit software measurement parameters. This may be of either a linear or one of various logarithmic scales, depending on measurement mode and settings. Further details of the analogue output to current conversion equations may be found in the QMG 422 manual. The length of time spent on the measurement channel is determined by two adjustable parameters in the Quadstar 32-bit software: *p-cal* and dwell. The *p-cal* value, typically 1.0, determines how long the channel is set before the mass ion current signal is averaged. This time is necessary in order to allow the amplifier to rise, overshoot and stabilise before the mass ion current is averaged. The dwell time then corresponds to the time over which the mass ion current signal is averaged before switching to another measurement channel. Although not clear from the QMG 422 manual, the total time for each channel is the *p-cal* value multiplied by the *dwell time*, plus the *dwell time*. In the DEMS Labview software the VI is capable of measuring either the *elm* or *mon* channel. In most cases where the analogue output from the EP 422 amplifier is linear, then *elm* is the preferred where the QMS signal is averaged in the Labview software. The part of the signal being averaged can therefore be assessed and the QMS settings easily optimised.

EP 422 Amplifier

The EP 422 electrometer amplifies the mass ion current signal output from the SEM (or faraday cup depending on QMS mode) and converts the signal into a proportional voltage between 0 and 10 V. This signal is then filtered, treated and averaged by the QC 422 bus (according to the Quadstar 32-bit filter parameter) in the QMG 422 unit which communicates the measurement value to the Quadstar 32-bit software. Despite that the Quadstar 32-bit software possesses options for eight amplification ranges, the EC 422 circuitry in fact only possess four (physical) amplification ranges, which should be considered in the optimisation of the DEMS instrument. These are noted in Table 3.4.

Much of the QMS signal noise appears (from testing) to arise from the EP 422 amplifier rather than the ion source or SEM. Ideally, therefore, the signal arising from the SEM is as large as possible so that the amplifier range can be set to a lower amplification range to minimise the measurement signal noise. This, however, is not always possible at reasonable SEM voltages, however, in general it is preferred that the SEM is adjusted so that the MID current is in the upper decade of the amplifier output range i.e. in the 10^{-7} amplifier range, the mass ion current signal should ideally be in the 10^{-7} region (1–10 V) rather than the lower 10^{-8} region (0–1 V) for optimum signal-to-noise ratio.

3.6.1.3 DEMS Measurement Software

A description of the software related to the DEMS experimental setup, namely the potentiostat, syringe pump, vacuum gauges and QMS will be presented in the following sub-sections. The majority of this software, with the exception of the commercial Quadstar 32-bit program, are Labview based programs designed or developed as part of this thesis for the DEMS instrument. Through thoughtfully designed Labview based software, the functionality, usability and productivity of any experimental setup can be considerably enhanced. Consequently, in order to aid experimentation using the DEMS instrument, additional functionality was incorporated into two previously written in-house software applications, the CV and Macro programs that were originally developed for RDE and MWE methodologies [15]. The descriptions of the custom Labview based software architectures and virtual instruments (VIs) presented here are intended to aid the understanding and interpretation of the graphical code. The commercial Quadstar software which controls the QMS, however, will be addressed first .

Quadstar 32-bit

The Quadstar 32-bit software is a modular application for QMS models 422/421 and QMS 200 Prisma^{TM}. The standard software is capable of performing qualitative and quantitative analyses and is user programmable. There are a number of programs that make up the Quadstar 32-bit software, however, most commonly used are the *Parset* (Parameter Setup), *TuneUp*, *Measure*, and *Dispsav* (Display Save) applications. During DEMS operation the Quadstar 32-bit *measure* program is launched either manually by the user or by the CV or Macro program during initialisation depending on the measurement mode. In both cases the QMS is operated in *MID Versus Time* mode and runs using a pre-defined *Measure MID parameter* (*.mip) file created in the *Parset* program. For more information on the Quadstar 32-bit software refer to the software documentation [21].

CV3

The CV3 program is a custom developed Labview program designed to perform CV electroanalytical experiments with DEMS. The software is an extension of the previous version (CV2) which was originally designed for single working electrode (SWE) and MWE electrochemical setups [15]. In CV3, the additional functionality which is required for the QMS component of the DEMS instrument was incorporated to aid DEMS experimentation; however, the control of the electrochemical experimentation essentially remains unchanged. The CV QMS software graphical user interface (GUI) is given in Fig. 3.24.

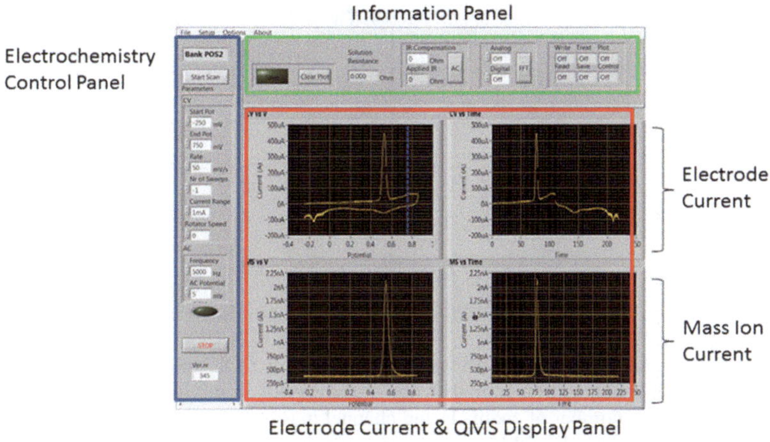

Fig. 3.24 Screenshot of the CV3 software front panel developed for the DEMS instrument. The CV3 software incorporates DEMS functionality combining CV electroanalytical experimentation with the acquisition, synchronisation, display and save of QMS mass ion current data in real-time

In the DEMS operation mode the faradaic electrode current and QMS data are measured, synchronised, displayed and saved in real time. The CV3 software control panel allows users to manually control the CV electroanalytical parameters such as potential range, scan rate etc. in real-time during an electrochemical experiment. In DEMS operation, QMS data is also measured, synchronised with the electrochemical and displayed along with the electrochemical data. In contrast to the standard SWE and MWE modes, the experimental data is display as both a function of the WE potential, and time allowing easier interpretation of the QMS mass ion currents.

The QMS data acquisition (referred to as DEMS mode) can be setup in the CV3 (and Macro or DC software described later) using a sub-panel whose user interface is given in Fig. 3.25. Here QMS data acquisition method must be defined (whether DDE or analogue DAQ) and the preferred QMS channel to display (if more than one).

Of particular importance here are the analogue DAQ settings, which define how the analogue signal from the QMG 44 unit is acquired, filtered, averaged and converted into a mass ion current. The settings here should correspond to the analogue output, amplifier range defined in QMS measurement parameters if the Quadstar 32-bit software otherwise the calculated mass ion current values will be incorrect. The settings for the DDE communication on the other hand will usually not require changing. The method in which data is obtained by the Labview QMS read VIs is described in Sect. 3.6.3.

The CV3 allows comprehensive user control over CV experimentation in real-time for SWE, MWE and DEMS apparatus; however, for more meticulous control of the electrochemical parameters and the use of other electroanalytical techniques such as chronoamperometry, the Macro software is preferred.

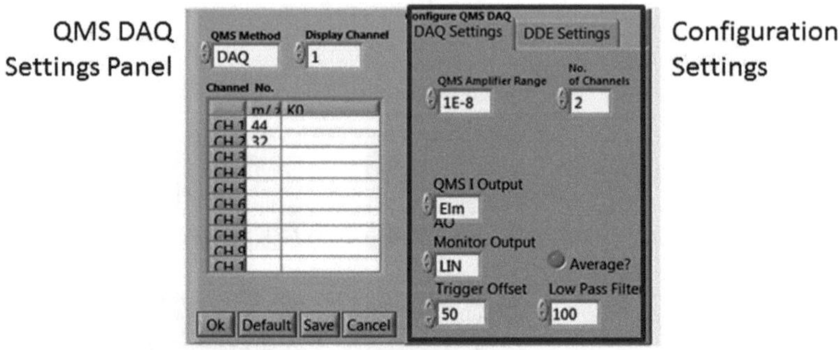

Fig. 3.25 Screenshot of the QMS settings control panel for DDE and DAQ. The pop-up panel allows the QMS data acquisition settings to be defined prior to the measurement. Of particular importance is that the number of channels and channel mass labels (m/z) are defined, and that all DAQ settings correlate to the QMS parameters defined in the Quadstar 32-bit software when using the QMS DAQ method

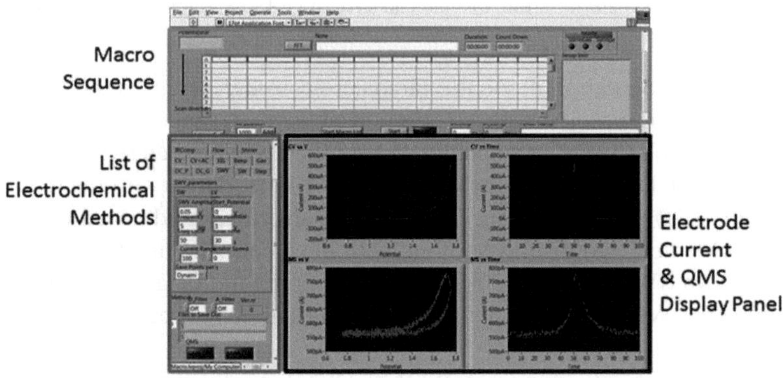

Fig. 3.26 Screenshot of the Macro software front panel further developed to incorporate the required DEMS instrument functionality. The Macro software allows experimenters to define a sequence of electroanalytical measurements (such as DC and CV), or parameter changes (e.g. flow rate) allowing meticulous control of experimentation

Macro

The custom Macro electrochemistry software program allows users to employ a wide range of electroanalytical techniques to study the electrochemistry of a reaction system. The software allows users to pre-define and execute a measurement sequence of electroanalytical techniques programmatically allowing meticulous control over electrochemical experimentation. The GUI of the Macro software is shown in Fig. 3.26.

The user first chooses from a list of electroanalytical techniques located to the lower left of the GUI panel, defines the measurement parameters and adds this to the Macro sequence list located at the top of the front panel. Once a list of electroanalytical methods has been defined, the user can then execute the sequence

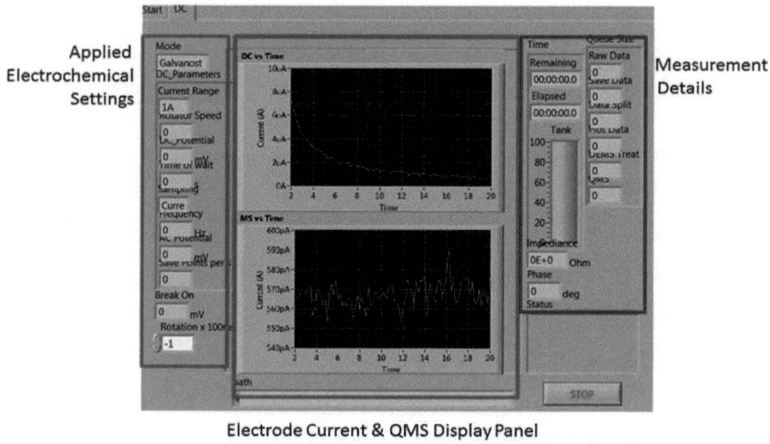

Fig. 3.27 Image of the DC3 software front panel developed for the DEMS instrument. The necessary QMS functionality required for the DEMS instrument was incorporated into the DC3 software, allowing chronoamperometric and chronopotentiometric DEMS measurements

and the measurements are performed. This powerful software was adapted to incorporate the CV3 application described previously, the DC3 and Syringe Pump Flow control VI's discussed in the following subsections to provide the necessary DEMS functionality. The DEMS functionality was not incorporated into all electroanalytical methods such as fast square wave voltammetry, because they are of limited practical use in DEMS. However, the architecture of the QMS measurement incorporation in the CV3 and DC3 VI's should provide a template to incorporate the QMS in other electroanalytical methods if and when required.

DC3

The DC3 software is designed to perform either chronoamperometric or chronopotentiometric electrochemical experimentation. The user interface of the DC software is depicted in Fig. 3.27.

Users can define the electrochemical parameters i.e. WE potential and time. The software is, however, more commonly used in the Macro software where a sequence of DC3 program executions can be defined, for instance in potential step measurements.

Syringe Pump Control

The syringe pump control software is used to control the KDS 200 syringe pump via the computer rather than manual interface. The GUI for the stand alone flow control application is given in Fig. 3.28.

The communication between the syringe pump and program is achieved via a R232 interface and must first be initiated. Details of the operational mode, syringe pump dimensions, volume and the desired flow rate must then be communicated to the syringe pump. Once this is achieved, the program can then control the

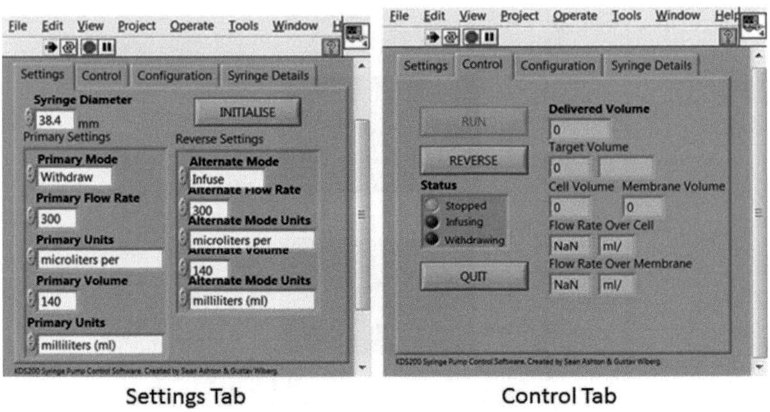

Fig. 3.28 Image of the KDS 200 syringe pump settings (*left*) and control panel (*right*) developed for the DEMS instrument. The functionality of this software was incorporated into the Macro software allowing the programmatic control of the syringe pump which is often desirable in DEMS experimentation

operation and direction of the syringe pumping. In this form, however, the program is of limited use simply because it is quite easy to make the same changes manually via the syringe pump interface. However, programmatic control of the flow rate is far more useful when incorporated into the Macro software in order to start or stop the pump, and manipulate the electrolyte flow during the course of an electrochemical experiment. It also eases the documentation of the experimental parameters which can be written into the DEMS measurement data file.

These electrochemistry software tools combined allow comprehensive control over the DEMS instrument for common measurement tasks. However, another measurement setup is required in order to perform calibration procedures required for the QMS.

3.6.2 QMS Calibration Setup

An overview of the measurement setup for the simultaneous acquisition of QMS and pressure data used for the calibration procedure described later in Chap. 4 is depicted in Fig. 3.29.

In order for the QMS to be calibrated, the pressure of a calibration volume and the mass ion current of the relevant species must be measured accurately and perfectly synchronised. In the QMS calibration measurement setup, pressure data may be acquired from the vacuum gauge control units using either or both a DAQ card and RS232 data communication. The DAQ card is used to acquire the raw analogue signal voltage output of the pressure gauge, whilst RS232 communication are averaged values provided by the Maxi Gauge unit. In general, the DAQ of

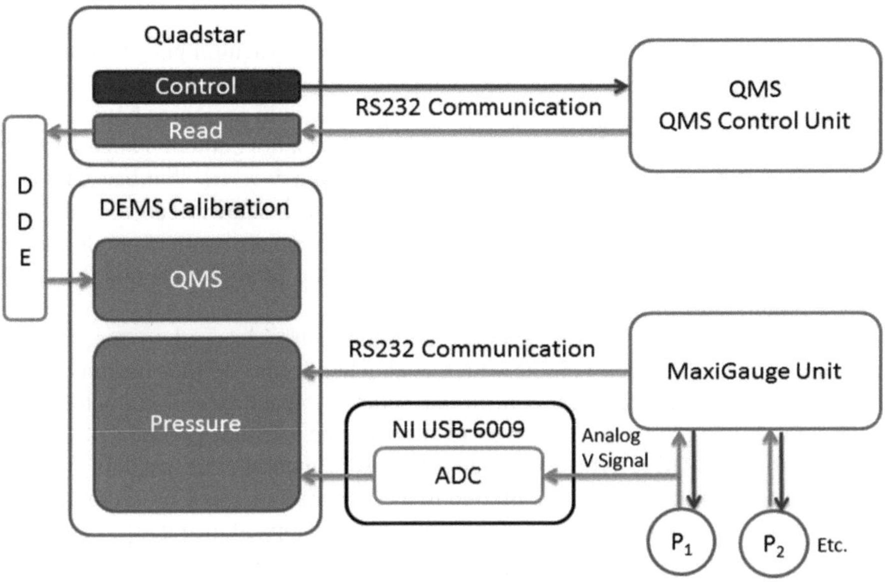

Fig. 3.29 Overview of the QMS calibration measurement hardware and software setup. The calibration procedure requires the precise measurement and correlation between the QMS mass ion currents and the pressure of a calibration volume (and vacuum system)

the analogue signal is preferred for the pressure measurement of the calibration volume. In the DEMS calibration software, however, the QMS data should be acquired using DDE communication between Labview and Quadstar software. In this case DDE communication is necessary simply because the DAQ card that is usually used to acquire QMS data during DEMS operation is now used to measure the vacuum gauge pressures.

3.6.2.1 Vacuum System Hardware

The DEMS vacuum pressure is monitored using a variety of vacuum gauges, namely two pirani gauges, two cold cathode gauges and a hot cathode gauge. An additional capacitance gauge is also used for the calibration of the QMS. Each of these gauges were purchased from Pfeiffer Vacuum, and are all powered and controlled using the Maxi Gauge control unit. There is of course additional vacuum hardware such as the turbo and rotary vane pumps (including their control units), however, these are neither especially adapted nor specific to DEMS. For a comprehensive overview of vacuum technology the reader is instead directed to Ref. [22–26].

Vacuum Gauges

The DEMS vacuum system employs five permanent vacuum gauges to monitor the vacuum pressure. The 1st and 2nd pumping stages employ a IKR 261 compact cold cathode gauge which is able to monitor high vacuum pressures between 10^{-2} and 2×10^{-9} mbar. The 3rd pumping stage employs a hot cathode gauge which is able to monitor vacuum pressures between 1,100 and 5×10^{-10} mbar. The cold cathode gauge could not be used on the 3rd pumping stage because of the disruptive influence of the magnet on QMS. Two pirani gauges are meanwhile used to measure the vacuum pressure between the back of the turbo pumps and the rotary vane (roughening) pumps.

For the calibration of the QMS an addition gauge is used to measure the pressure of the calibration volume which is attached to the DEMS vacuum system in place of the electrochemical cell. In this setup, a capacitance manometer gauge with ceramic diagram was employed to monitor the vacuum pressure of the calibration volume. These types of gauge are able to measure the vacuum pressure very precisely, although they are limited to a comparably narrow pressure range of $\sim$2 decades. The chosen gauge for calibration was a capacitance gauge capable of monitoring vacuum pressure between 1 and 110 mbar. Each of the vacuum gauges require the MaxiGauge vacuum measurement and control unit to operate.

MaxiGauge

The MaxiGauge control unit is required to operate and monitor the six Pfeiffer Vacuum pressure gauges used by the DEMS instrument. When operated the MaxiGauge first identifies the attached vacuum gauge, provides the necessary power for operation, and then monitors the pressure gauge measurement voltage output which it averages and converts into a pressure value that is then displayed on the MaxiGauge display panel. This allows users to observe the vacuum system pressure which is often all that is desired.

For the QMS calibration procedure described later in Chap. 4 Sect. 4.5, it is necessary to accurately measure the pressure decay of a calibration volume and precisely correlate this with a corresponding QMS mass ion current. In order to do this, however, we once again require a data acquisition setup to record the vacuum pressure. Pressure measurements can be acquired from the MaxiGauge unit via a R232 interface and programmatic communication between the device and Labview software. This is an acceptable solution when monitoring gradual changes in vacuum pressure where time resolution is not critical because the DAQ rate is relatively slow, and irregular due to the nature of RS232 communication. This is problematic for the QMS calibration procedure where the pressure changes relatively quickly and the measurement is very time critical. A faster alternate method of acquiring the vacuum pressures with precise time resolution is to measure the original pressure voltage output arising from the pressure gauge using a DAQ card. The vacuum gauge manufacturer, however, does not make this straightforward because the MaxiGauge unit does not possess any analogue signal outputs, and the nature of modern vacuum gauges does not make it easy to probe the signal directly. However, by construction a special adapter cable which lies between the

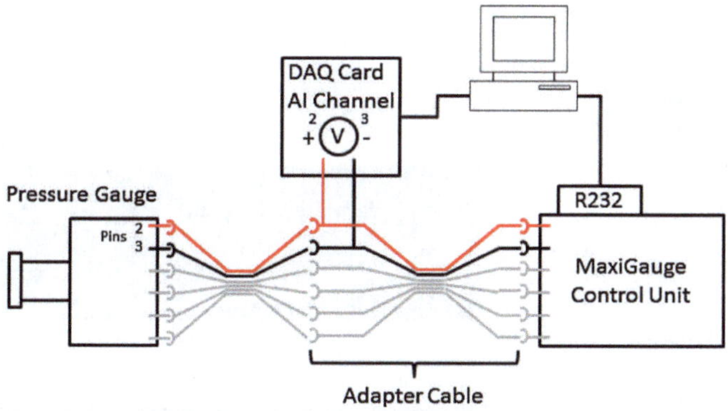

Fig. 3.30 Schematic of the analogue pressure gauge measurement system used for QMS calibration. An adapter cable was created to directly probe the pressure dependant potential output of the pressure gauge to precisely acquire the analogue signal using the DAQ card rather than via the averaged measurement obtained through the MaxiGauge (which is not time resolved)

pressure gauge and control unit, the vacuum gauge measurement output may be probed without impinging on the vacuum gauge operation. The adapter cable therefore lies between the connection between the pressure gauge and Maxi Gauge unit, as depicted in Fig. 3.30.

The vacuum measurement voltage output can therefore be measured directly and converted into the corresponding pressure via a certain equation provided in the gauge operating manuals (see Ref. [22–24]). For the capacitance gauge in the QMS calibration setup the output between 0.1 and 10 V simply corresponding to a pressure range of 1–100 mbar. For more details of the MaxiGauge operation refer to the operating manual [25]. The acquisition and synchronisation of the QMS and vacuum pressure meanwhile is achieved using the QMS calibration software.

3.6.2.2 QMS Calibration Software

The QMS calibration software was programmed in order to facilitate the calibration of the QMS and avoid the tedious and time consuming nature of the calibration procedure which in some fields may often be avoided unless absolutely necessary for a certain analysis. The calibration software is used to measure, synchronise, display and record the QMS mass ion current and calibration vacuum gauge pressure simultaneously. Once a QMS calibration measurement has been performed, a data treatment panel is also included in the software which is capable of importing and treating the calibration measurement data to extract the calibration constant. The DEMS (QMS) calibration software GUI is shown in Fig. 3.31.

In order to calibrate the QMS instrument, the pressure of a calibration volume and DEMS vacuum chamber must be recorded in parallel to the mass ion currents

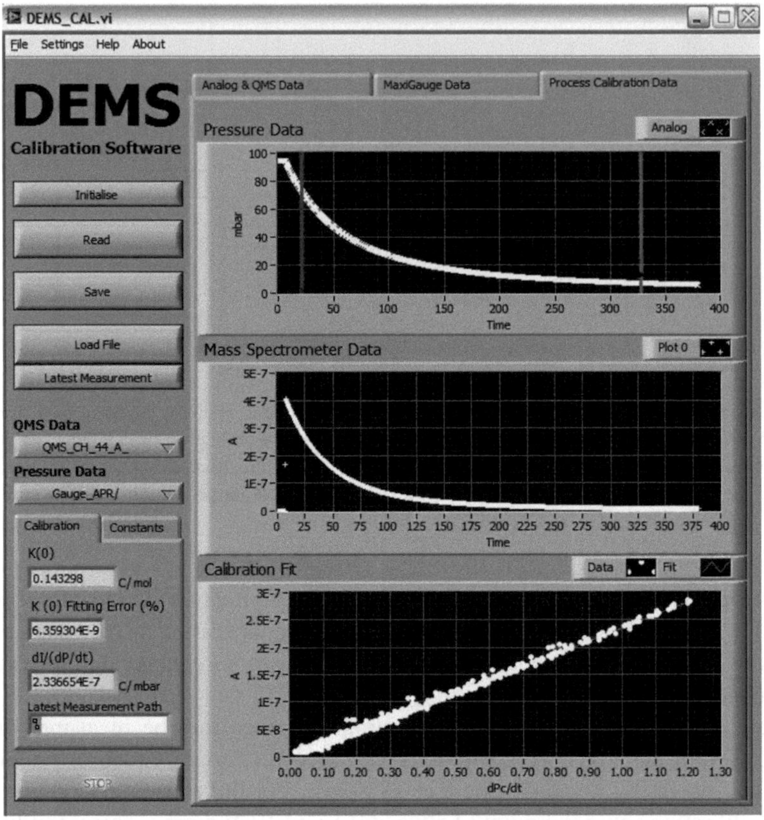

Fig. 3.31 Image of the DEMS (QMS) calibration software front panel developed for the DEMS instrument. The DEMS (QMS) calibration software is capable of both measuring, displaying and recording pressure and QMS mass ion current data in real-time. Following a calibration measurement procedure, the software is also imports and treats the calibration data to provide the calibration constant, K^0

relevant to the species of interest. Once initialised and started the software begins to read, synchronise and display the QMS and pressure gauge data. When the calibration measurement is ready to be performed, the user can then chooses to begin to save the data points which are written to a text (.txt) file in a tab delimited format. The calibration procedure can then performed and once finished, the user can then stop the saving of measurement data points. The calibration procedure can then be repeated to ensure reproducibility of the measurement, and then stopped. The QMS calibration files are then saved to a user defined folder name organised in a folder structure according to the measurement date. Further details of the calibration setup and measurement methodology are presented in Chap. 4 Sect. 4.5.

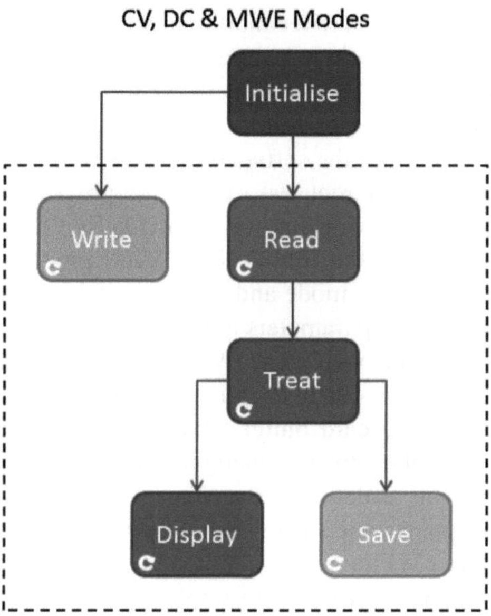

Fig. 3.32 A diagram of the electrochemistry Labview software conceptual architecture in standard CV, DC and MWE operational modes. Following the initialisation, the software possesses a write VI to control the potentiostat whilst the read measures the potentiostat response. The electrochemical data is then passed to the treat VI where it is treated before it is displayed (in the GUI) and saved. By passing the data between the VIs using Labview Queues, the VIs are able to operate at different execution speeds allowing easy synchronisation of the various processes

3.6.3 Labview Software Architecture

The purpose of this section is to present an overview of the architecture of the custom Labview programmed electrochemistry software (CV3), and illustrate how the required DEMS functionality has been incorporated into the software. Rather than attempting to document and explain the graphical code which would take a thesis in itself (and would not be appreciated) the conceptual architecture of the code will be described from a top level perspective. This should at least aid in the understanding of the principle structure of the Labview code to those with little or no programming experience, whilst assisting an experienced Labview programmer to interpret the structure of graphical code. The architecture of the CV3 (and DC3) software will first be presented in the standard single working electrode (SWE) and multiple working electrode (MWE) modes. This is then followed by an explanation of how required DEMS functionality was incorporated into the software. The architecture presented here can in principle be used for many programming solutions.

In Labview based graphical programming it is good practice to encapsulate segments of code with a specific function or task within a Labview virtual instrument (VI) in order to ease the readability and management of the code from a higher programming level. The top level architecture of the CV3 and DC3 Labview VIs, when operating in either SWE or MWE mode (i.e. no DEMS functionality) is conceptualised in Fig. 3.32.

In the SWE or MWE modes the software relies on the execution of six VIs that are each responsible for a different task. When the program is started an initialisation VI(s) executes once (during start or restart) before five VIs execute continuously. In the CV3 software the initialisation VI(s) prepare the DAQ card and obtains the setup, measurement mode and experimental parameters defined by the user in the GUI. The relevant parameters are then passed onto the Write and Read VIs which execute simultaneously. The Write VI is responsible for writing the potential perturbation values (such as for a potential ramp/waveform or fixed potential value) to the DAQ card buffer which is transformed into an analogue voltage output that is applied to the analogue potentiostat WE potential control input. At the same instant, as defined by a trigger, the Read VI begins to read the DAQ analogue input signals which arise from the potentiostat measurement voltage outputs i.e. the measured WE current and potential. This data is then passed to the Treat VI which transforms the data, for example, by converting the current voltage measurements into the corresponding current, filtering and averaging the raw measurement data. In the CV3 software, this Treat VI is designed to average the electrochemistry measurement data to a give single data point per mV (or per second in DC mode) independent of the potential scan rate. The Treat VI then outputs the treated data to the Display, and a Save VIs. The Display VI plots the measured data in the appropriate format i.e. either current vs. potential (CV3), or current and potential versus time (DC3). The Save VI meanwhile writes the measured data to a tab delimited text file (.txt). The synchronisation and timing of these VIs is aided considerably using a set of built-in Labview Queue VIs. The Queue VIs allows data to be passed and buffered between each of the five VIs, allowing each VI to execute at different speeds. By furthermore placing the VIs in separate CPU threads, and assigning the VIs and execution priority, the execution of time critical VIs such as the Write and Read VIs can be prioritised over the Treat, Display or Save VIs which are not time critical and can execute at slower rates, or with a slight delay without losing measurement data. Further details of the aspects of the Labview program in SWE and MWE modes may be found in Ref. [15].

In order to reduce management of the Labview programming and utilise the already powerful existing CV2 and Macro software for SWE and MWE methods, the required DEMS functionality was incorporated into the existing CV2 software architecture, rather than construct a new separate piece of software. In order to incorporate the required functionality of DEMS in the CV3, DC3 and Macro software, two additional VIs, the QMS Read and DEMS Treat VIs were introduced, whilst the Save and Display VIs were modified to offer more flexibility (i.e. the same core VI's can be used in SWE, MWE and DEMS modes). The conceptual architecture of the electrochemistry software in DEMS operation mode is presented in Fig. 3.33.

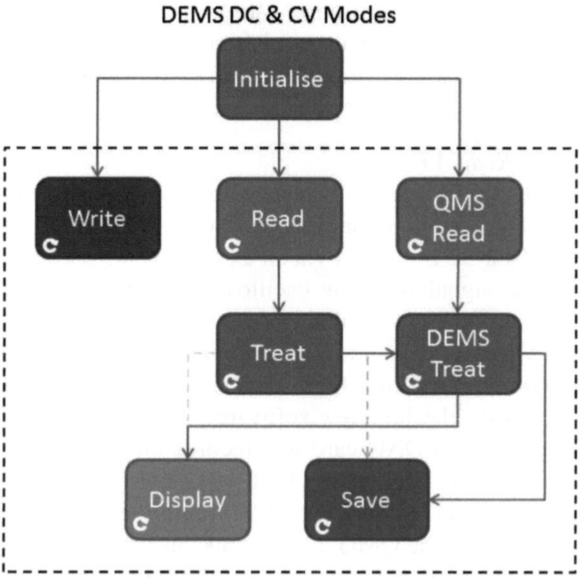

Fig. 3.33 Diagram of the electrochemistry Labview software conceptual architecture in DEMS operation mode. In this mode the Queue from the Treat VI to the Display (normally used in CV/MWE modes) is cancelled whilst the Queue to Save is diverted to the DEMS Treat VI. This VI synchronises QMS measurement data (from QMS Read) with the electrochemical data. The DEMS Treat then passes the combined electrochemical and QMS data to the Display and Save VIs.

The QMS Read VI is responsible for acquiring data from the QMS, which may be obtained using two modes: either analogue DAQ or DDE software communication, which are each later described in more detail. This VI then sends data via a Labview Queue VI to a DEMS Treat VI. In DEMS mode the functionality of the standard Treat VI is changed and instead of sending data directly to the Display and Save VIs, the data for the Display VI is cancelled whereas the data intended for the Save VI is re-directed via a queue to the DEMS Treat VI. The DEMS Treat VI then synchronises the electrochemical data with the QMS data in real-time before the new data is passed to the Display and Save VIs which plot and save the combined electrochemistry and QMS data in the desired format. By using this structure, the electrochemistry VIs are largely unchanged in the DEMS operation mode and we simply introduce two new VIs and re-direct the treated electrochemistry data. A more detailed description of the Labview VIs used in the DEMS software is described in the following subsections.

Initialise VI

The purpose of the QMS Initialise VI(s) is to perform the all the necessary functions that must be done prior to the start of the DEMS measurement. All the parameters that are relevant to the QMS initialisation and measurement are placed into Labview global parameter class called 'QMS_Config_lvclass'. This contains all details relevant to the VIs required for QMS VIs such as the QMS status, DAQ measurement or DDE service details. These parameters may be manipulated using the QMS settings dialogue box presented previously in Sect. 3.6.1.3. The QMS configuration settings are furthermore stored in the same 'ini.txt' file which retains details of the electrochemical setup parameters (such as potentiostat id, measurement mode etc.) when the software is closed. Once the measurement software is started the initialisation VI(s) execute and prepare the DEMS instrument for operation. In DDE data acquisition mode this initialisation procedure also involves launching the Quadstar 32-bit software, which is used more regularly by the QMS calibration software rather than CV3 etc.

QMS Read VI

The QMS Read VI is responsible for acquiring the QMS data points, which may be achieved in either of two ways: (1) by using DDE communication with Quadstar 32-bit, or (2) via an external DAQ card that reads the amplified mass ion current signal from the oscilloscope connections on the QMG 442 control unit. DDE is a technology that enables communication between software applications in the Microsoft© Windows operating system environment. In this case, DDE is used to communicate the QMS data points obtained by Quadstar 32-bit *measure* program with the Labview software VI. Alternatively, raw QMS data may be measured using a DAQ card that records the signal output from the analogue outputs (AOs) in the back of the QMG 422 unit. The DAQ card records the signal from *sync*, *scan* and either *elm*, the raw amplifier AO signal, or *mon*, the measured value averaged by the QMG 422 unit. Details of how the QMS data is extracted from the oscilloscope connection of the QMG 422 (the QC 442 unit) unit is described in Sect. 3.6.1.2 (see Fig. 3.25).

In general, data acquisition via the oscilloscope connection is favoured over DDE for a number of practical reasons. Firstly, DDE communication Quadstar must be initialised controlled and closed using the Labview VIs which locks any manual control of Quadstar 32-bit software program by the user when the CV3, DC3 or Macro software is operating. This is unsatisfactory because the Labview software program must be restarted each time the QMS settings are change, and furthermore DDE can use considerable computer resources. In contrast, using the DAQ card the Quadstar 32-bit measure software runs independently of the Labview VIs and QMS settings are easily adjusted when desired. An additional advantage of using the DAQ card, is that the raw electrometer signal may be assessed using the *elm* analogue output. The raw signal can then be treated using appropriate analogue signal filters and averaged using Labview rather than (blindly) relying on the QMG 422 units and Quadstar 32-bit DDE or *mon* oscilloscope output which is less flexible and more of a 'black box'. In cases where unusual QMS settings are used the QMG 422 unit data output can deviate significantly from the actual values, particularly when inappropriate filters, p-cal values or QMS parameters are applied. Finally, DDE generally requires more computer resources than the DAQ method, which not only limits the ability of the computer to perform other tasks but there is also a risk of time critical data points to arrive late or be lost. Regardless of the method used to acquire QMS data, once obtained the data points are sent to the DEMS Treat VI.

DEMS Treat VI

The DEMS Treat VI is responsible for synchronising the QMS data points with the electrochemistry data points, a process which is crucial to the precision of the instrument. However, combining data from two separate sources that measure at different sampling rates (which can both change depending on the measurement parameters) can be quite tricky, particularly when DAQ rates are fairly slow (such as with the QMS) in the desire that no measurement points are discarded. In the electrochemistry Treat VI, the measured WE current is treated to provide a single data point per mV, regardless of the potential scan rate of a CV, or a single data point per second for chronoamperometry (DC3). The data sampling frequency of the QMS is, however, comparably slow compared to the electrochemical measurement, and is furthermore sensitive to the QMS parameter settings, particularly in terms of the number of different mass ion current channels measured, and their corresponding p-cal and dwell times. Owing to the relatively slow sampling rate of the QMS, the QMS data cannot simply be averaged down to provide a single data point per mV (unless a single mass ion current is measured in the QMS DAQ mode, however, this is generally not the case).

The time, current and potential data from the electrochemistry Treat VI which is sent to the DEMS Treat VI has been averaged to provide a single data point per mV (or s^{-1} in the DC3) independent of the potential scan rate. The QMS data point frequency is, however, comparably slow and cannot be simply averaged down to a single point per mV, particularly for scan rates >ca. 5 mV s^{-1}. In order to allow some flexibility in terms of the possible CV scan rates, along with QMS settings (e.g. dwell times) the DEMS Treat VI synchronises the QMS data with the electrochemistry data differently for cases where the QMS sampling frequency is faster than a single data point per mV, from cases where it is slower. The manner used in the DEMS Treat VI to synchronise the QMS data with the electrochemical data is illustrated in Fig. 3.34. For situations where the QMS sampling rate is faster than a single data point per mV, the QMS data points are simply averaged. This may be the case when few QMS channels are measured and/or and the potential scan rate of the CV is slow. More commonly, however, the QMS data point frequency is slower than one point per mV and the QMS data is retained rather than averaging the faradic current data down to the QMS sampling frequency (this way we do not lose any of the electrochemical data). Another approach could have been to extrapolate QMS points where a data point was missing, however, this was considered highly inadvisable (one would lose track of the true measurement data).

Overall, the DEMS Treat VI allows a flexible degree of variation in electrochemistry scan rates, and QMS sampling frequency without loss of electrochemistry data. The duplication of the QMS data points, however, necessitates the removal of the duplicate data points during post measurement data treatment, particularly where smoothing of the QMS data is desired (this removal is performed using the DEMS data analysis program described later).

Plot VI

The QMS plot VI displays the electrochemical and QMS data that was synchronised in the QMS treat VI. In contrast to the CV plot, the electrochemical and QMS data is plotted as both a function of both potential and time. This way the QMS mass ion currents are more easily interpreted as a function of time, compared

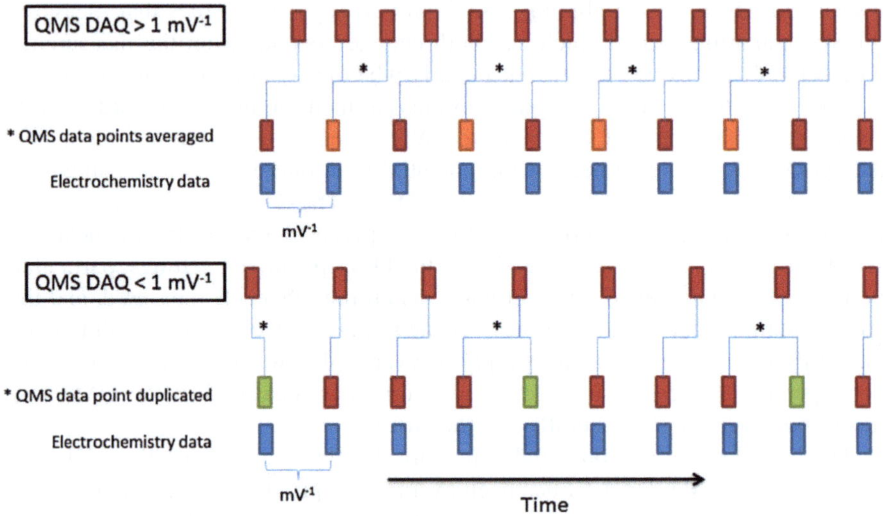

Fig. 3.34 A diagram illustrating how the QMS data points are correlated to the electrochemistry data. When the QMS DAQ speed is faster than 1 mV s^{-1}, the QMS data points are averaged, whereas when it is slower, the QMS data points are duplicated when a new QMS data point is unavailable. This allows a flexible range of scan rates to be used without needing to endlessly adjust QMS parameters, or losing electrochemical measurement data

to the electrochemistry which is best analysed as a function of the electrode potential. The Display VI also retains data from the previous three CVs in the display of the latest CV so that the experimenter can observe changes in the electrochemistry of the WE, for instance when changing electrochemical parameters or assessing stability and reproducibility of a system. In the CV3 and DC3 programs the display VI incorporates the appropriate plots for SWE, MWE and DEMS techniques rather than using separate VIs making the Labview code more manageable.

Save VI

The Save VI is responsible for saving the DEMS measurement data in '.txt' text file format. The Save VI was adapted so that the VI would work in more general cases, i.e. it functions for each of the SWE, MWE and DEMS modes of the software. The DEMS data file produced is a tab delimited text file, as shown in Fig. 3.35 containing details of the experimental parameters (either auto assigned or defined in the CV/Macro by the user), electrochemistry data (time, WE potential and WE current) and in DEMS mode, the QMS data.

A new DEMS data file is created for each CV or chronoamperometric measurement, and is stored in a user defined folder name that is organised in a folder hierarchy according to date, month and year. The tab delimited text file format was chosen because it is easily read and can be accessed or imported into the majority of data analysis software.

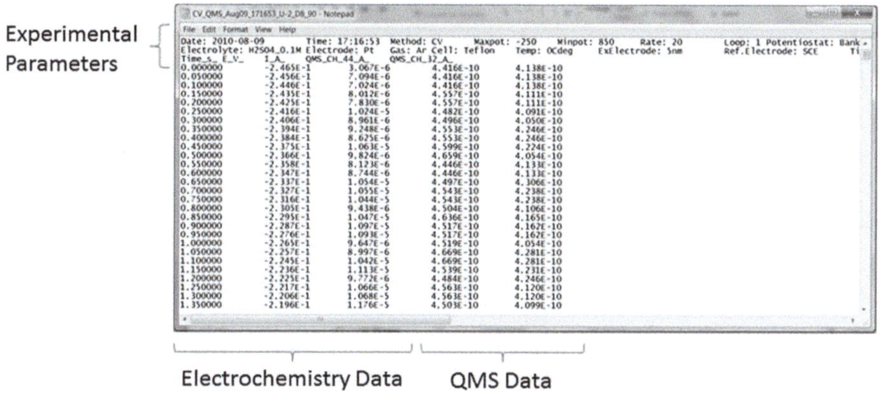

Fig. 3.35 Screenshot of the DEMS save data file indicating the file format. The experimental parameters relevant to the electrochemical measurement are stored in the file header. The electrochemistry and QMS mass ion current channel data are then written in tab delimited columns. This allows experimental data to be easily imported into common data analysis software such as *Excel* or *Igor Pro*, or better: the custom Labview programmed DEMS analysis software

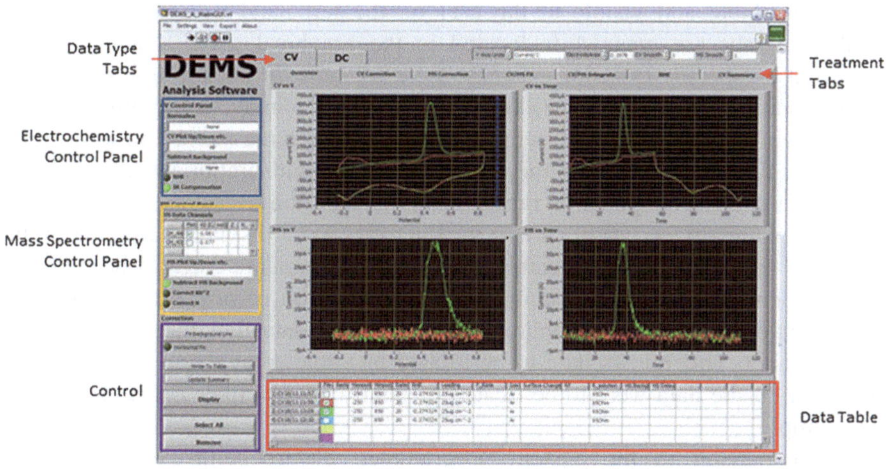

Fig. 3.36 Screenshot of the DEMS analysis software front panel developed as part of this thesis. The DEMS analysis software allows common, necessary data analysis and treatment tasks that are specific to DEMS to be performed quickly and easily. The software also allows treated data to be exported to other analysis programs such as *Excel* or *Igor Pro*

3.7 Data Analysis

Data treatment can be very time consuming and in situations where tasks are repetitive, specifically designed data treatment software can greatly enhance productivity. Consequently, a custom Labview software program was also

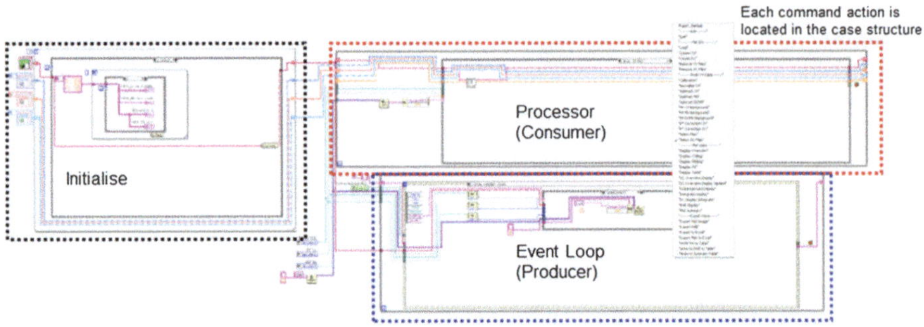

Fig. 3.37 Screenshot of the queued state machine producer–consumer architecture of the DEMS Analysis software. The event loop is responsible for the programmatic response to user events (*producer*), whereas the processor (*consumer*) executes the corresponding code. This architecture is a very useful method of solving many programming tasks

developed in order to perform common measurement data treatment tasks that are specific to DEMS. The DEMS Analysis software GUI is depicted in Fig. 3.36 .

The GUI is arranged so that common display or data treatment options are always available and visible to the user on the left hand side of the panel. A number of tabs are meanwhile located in the graphical view, which are each organised by specific task, for instance, there are tabs for background subtraction and integration. The DEMS measurement data.txt files may be imported into the program via the runtime header which are then added to the interactive table at the bottom of the GUI displaying the file names along with important measurement parameters such as the potential range, scan rate, solution resistance etc. Using this table, users are able to easily choose the data files to be displayed and treat in the necessary manner. The data treatment tasks for the faradaic current and mass ion current such as, background subtraction, normalisation or post measurement *iR*-compensation etc. are then applied on the selected data files and plotted in the display panel. This allows data to be treated in a matter of a few minutes rather than hours if the data were to be treated manually, using either *Excel* or *Igor Pro* for example.

In comparison to the custom programmed electrochemistry measurement software, the graphical code of the DEMS data analysis software is comparatively more straightforward, with a (hopefully) clearer structure. The graphical code of the software will therefore be discussed more specifically, in order to direct future users or programmers of the software to adding additional functionality to the DEMS analysis software if and when desired for special data treatment tasks. An annotated image of the top level graphical code of the DEMS data analysis program is shown in Fig. 3.37. The program essentially possesses three loops: an *initialise loop* (which executes when the program starts), *event* (or producer) and *processer* (or consumer) *loops* which execute depending on the user input from the GUI.

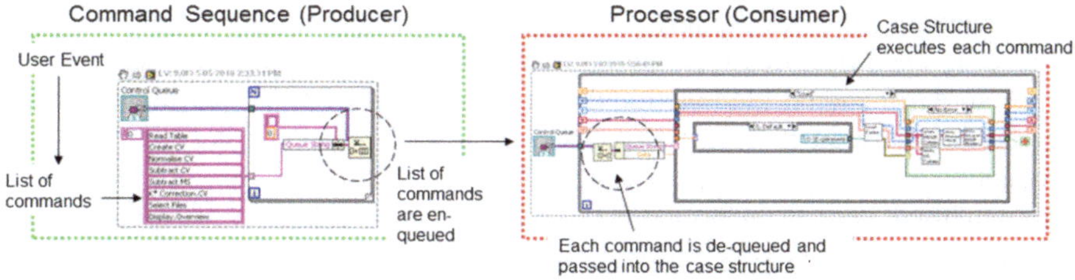

Fig. 3.38 Diagram representing the structure of the data treatment command sequence (*producer*) and sequence processor (*consumer*) of the DEMS analysis software. The event loop produces an array of strings which contain the tasks necessary to execute the user event. This list of commands is then executed on the imported electrochemical data, treating the data in the desired way and displaying the result

The *initialise loop* only executes once when the Labview program is started, and is responsible for resetting the GUI and writing the necessary front panel references to a class VI. The event and processor loop then execute continuously. The *event loop* handles all user inputs by recognising an event, such as a user mouse click on a GUI button which corresponds to a certain task such as import, treat or display data. The event loop then executes and produces a command (or set of commands) which is passed to the *processor loop*. The processor loop has two main functions: (1) it imports and stores the raw DEMS data in the *shift register* of the processors *while loop* in a file *data class* and (2) it performs all the necessary data treatment tasks. Although these tasks will be specific to the data treatment being performed, it generally involves creating an interpolated duplicate of the raw DEMS data, which is then treated in the desired manner before being plotted in the DEMS analysis display. The data treatment tasks are, however, determined by the user and in turn the *event loop*. The communication between the *event loop* and the *processor loop* is handled using the Labview Queue VIs.

When the user executes a certain event such as *display data*, the *event loop* produces a list of commands (or arrays of *strings*) which must be performed to display that data. This array of *strings* is then enqueued into the Labview *queue* in the *event loop* (producing commands), which are then *de-queued* in the corresponding order in the processor (which consumes the commands). Each of these commands (or *strings*) correspond to a case in the processor while loop case structure. The processor then *de-queues* each *string* and executes the corresponding case in the processor *case structure* for each loop iteration. The key elements of this graphical code are depicted in Fig. 3.38. These array of *strings* typically involves reading the GUI table (to read the data table values), create the CVs (which duplicates and interpolates the raw data into a treat data *class*), performing a series of data treatment steps (to treat the treat data *class*), selecting the data (which obtains the chosen files to be displayed) and finally, displaying the treated data.

This structure should allow the easy addition of new data treatment tasks by simply creating a new case in the processor case structure, and adding the corresponding *string* to the array of *strings*, which may be desired in the future, depending upon data treatment tasks. Furthermore, the software also possesses an export to *Excel* and.txt file function that allows treated DEMS experimental data to be further analysed in either *Excel* or *Igor Pro* for special analysis or for the production of publication quality figures. The structure of the DEMS analysis program used here is in principle a *queued state machine producer consumer architecture* which is very powerful and can be used to solve a wide range of programming tasks, further details of which may be found in Ref. [26].

3.8 Summary

The aim of this chapter was to present details of the DEMS instrument design, along with a description of the required hardware and software that were developed in order for the instrument to not only be functional, but also very usable for experimentation.

An electrochemical half-cell was designed that is in principle, of very close geometry to the dual thin-layer DEMS flow cell design presented previously in Chap. 2. This allows electrochemical experimentation on bulk and thin-film electrodes but requires a flow of electrolyte. This was achieved by withdrawing electrolyte from an electrolyte supply reservoir using a precision syringe pump. In this particular cell construction a certain amount of flexibility was incorporated into this design to, for instance, allow the exchange of individual cell parts such as the flow block to give a different flow regime over the WE, and is considered rather intuitive to the cell development processes. Furthermore, the electrochemical half-cell construction possesses an all-Teflon or Kel-f flow pathway for the aqueous electrolyte, allowing alkaline solutions to potentially be used without contamination from glass constituents.

The membrane interface consists of a microporous PTFE membrane which is mechanically supported by a stainless steel frit. The total flux of gas through the membrane when separating the aqueous electrolyte from the vacuum system is crucial to the vacuum system design, and obtaining the necessary vacuum pressure required to operate the QMS. It is also crucial to ensure that the membrane is sealed between the vacuum and the cell, preventing an additional leak of air into the vacuum system increasing the vacuum system pressure and background QMS mass ion currents. In general, the membrane interface compartment should possess a large geometric area to volume ratio in order to ensure high collection of the electrochemical reaction species.

The vacuum system was of a differentially pumped design, consisting of three vacuum chambers. The membrane interface is attached to a 1st pumping stage which operates at a rather high pressure of 10^{-2} mbar. This is attached to a 2nd pumping stage operating at $\sim 10^{-5}$ mbar containing the CB-ion source. The 1st and 2nd vacuum chambers are connected via tubular aperture which directs the flux of gas entering from 1st chamber through the CB-ion source of the QMS. This was to allow a higher pressure at the ion source in order to maximise the sensitivity of the QMS. Finally, a 3rd pumping stage also lowers the pressure of the chamber containing the SEM of the QMS to $<5.0 \times 10^{-6}$ mbar range required for long-term, reliable operation of the SEM.

The operation and overall usability of the DEMS instrument, however, relies on a number of additional commercial and custom developed hardware, and software produced as part of this thesis. The electrochemical setup employed a commercial BANKS Pos2 analogue potentiostat, a NI DAQ card and custom-built NID to control the electrochemistry and apply appropriate *iR*-compensation. The QMS and vacuum pressure gauges were meanwhile controlled via commercial hardware. Potentiostat control and measurement was achieved using NI DAQ card and custom Labview CV/DC and Macro software, originally developed for the RDE methodology. As part of this thesis, however, these software were further developed in order to incorporate the mass spectrometry aspects of the DEMS instrument. The resulting software with added DEMS functionality allows electrochemical data to be measured, synchronised, recorded and displayed alongside QMS measurement data in real-time. The acquisition of QMS data meanwhile was achieved using either DDE software communication, or via a NI-DAQ card to record the oscilloscope output of the QMS control unit. A separate hardware and software setup was also constructed using NI DAQ cards and in-house Labview programmed software to calibrate the QMS. Finally, a custom data analysis program was developed in order to perform data treatment tasks that are common in DEMS.

References

1. Mayrhofer KJJ et al (2009) An electrochemical cell configuration incorporating an ion conducting membrane separator between reference and working electrode. Int J Electrochem Sci 4(1):1–8
2. Jusys Z, Massong H, Baltruschat H (1999) A new approach for simultaneous DEMS and EQCM: electro-oxidation of adsorbed CO on Pt and Pt-Ru. J Electrochem Soc 146(3): 1093–1098
3. Baltruschat H (2004) Differential electrochemical mass spectrometry. J Am Soc Mass Spectrom 15(12):1693–1706
4. Holzbecher E et al (2006) Modellierung von Dünnschichtzellen (Thin Layer FLow Modelling). In: COMSOL Users Conference. 2006. Frankfurt
5. Fuhrmann J et al (2009) Numerical calculation of the limiting current for a cylindrical thin layer flow cell. Electrochim Acta 55(2):430–438
6. Wolter O, Heitbaum J (1984) Differential electrochemical mass-spectroscopy (Dems) - a new method for the study of electrode processes. Berichte der bunsen-gesellschaft-Phys Chem Chem Phys 88(1):2–6
7. Fujihira M, Noguchi T (1993) A novel differential electrochemical mass-spectrometer (DEMS) with a stationary gas-permeable electrode in a rotational flow produced by a rotating rod. J Electroanal Chem 347(1–2):457–463
8. Jusys Z, Behm RJ (2001) Methanol oxidation on a carbon-supported pt fuel cell catalysta kinetic and mechanistic study by differential electrochemical mass spectrometry. J Phys Chem B 105(44):10874–10883

9. Smith SPE, Casado-Rivera E, Abruna HD (2003) Application of differential electrochemical mass spectrometry to the electrocatalytic oxidation of formic acid at a modified Bi/Pt electrode surface. J Solid State Electrochem 7(9):582–587
10. Operating Instructions, Quadrupole mass spectrometer system, QMG 422. Vol. BG 805 983 BE: Pfeiffer Vacuum GmbH
11. Velghe M, Deacon DAG, Ortega JM (1984) Realization of a variable aperture diaphragm working in ultra high vacuum. Appl Opt 23(21):3851
12. Furman SA et al (2001) Improving the detection limit of a quadrupole mass spectrometer. J Vac Sci Technol A: Vac Surf Films 19(3):1032–1033
13. Gerlach W (1996) PPM analysis by using quadrupole mass spectrometers with closed ion source. Vacuum 47(4):371–374
14. Wiberg GKH (2010) The devlopment of a state-of-the-art experimental setup demonstrated by the investigation of fuel cell reactions in alkaline electrolyte, in Lehrstühl für Physicalische Chemie. Technische Universität München: München
15. KD Scientific Model 200 Series Manual: KD Scientific Inc
16. Britz D (1978) iR elimination in electrochemical cells. J Electroanal Chem 88(3): 309–352
17. Oelßner W, Berthold F, Guth U (2006) The iR drop—well-known but often underestimated in electrochemical polarization measurements and corrosion testing. Mater Corros 57(6): 455–466
18. Lamy C, Herrmann CC (1975) A new method for ohmic-drop compensation in potentiostatic circuits - stability and bandpass analysis, including effect of faradaic impedance. J Electroanal Chem 59(2):113–135
19. Quadstar 32-bit, Application Program for Quadrupole Mass Spectrometers QMS 422 and QMS 200.: Pfeiffer Vacuum GmbH
20. Homan F (1980) Design for an inexpensive externally operable variable aperture to be used in UHV. J Vac Sci Technol 17(2):664–664
21. Operating Instructions, Compact Capacitance Gauge: Pfeiffer Vacuum GmbH
22. Operating Instructions, Compact FullRange BA Gauge: Pfeiffer Vacuum GmbH
23. Operating Instructions, Compact Cold Cathode Gauge: Pfeiffer Vacuum GmbH
24. Operating manual, TPG 256 A, Vacuum measurement and control unit for Compact Gauges, MaxiGauge
25. Operating Instructions: TRIVAC B Rotary Vane Vacuum Pump D 16 B/D25 B: Leybold Vacuum
26. Lukindo A LabVIEW Queued State Machine Architecture. 2007 01 Oct 2007; Available from: http://expressionflow.com/2007/10/01/labview-queued-state-machine-architecture/

Chapter 4
Practical Aspects of the DEMS Instrument

This chapter concerns the more practical aspects of the DEMS instrument, covering the instrument characterisation, influence of experimental variables and measurement parameters, calibration procedures and finally, error and maintenance considerations. The characterisation and evaluation of the DEMS instrument is crucial to finding optimum experimental parameters, discovering possible areas of improvement during the development process, and to understanding the capabilities and constraints of the instrument before any research is conducted. This process is separated into the evaluation of the three principle DEMS components, to re-iterate the electrochemical cell, the membrane interface, and the vacuum system including QMS whose capabilities combine towards the overall performance of the instrument. Following this, the methodologies and setups used to optimise and calibrate the QMS and the DEMS instrument as a whole are presented. The contents of this chapter should serve to provide a practical understanding of the DEMS instrumental variables and document calibration procedures so that the instrument can be used in confidence by future users.

4.1 Electrochemical Cell

A thorough understanding of the DEMS electrochemical cell characteristics is necessary in order to differentiate experimental observations that purely arise from instrumental parameters or measurement methodology, from those that actually result from the electrochemistry of the WE. An immediately apparent and undesirable characteristic of the DEMS electrochemical flow cell is its inherent ability to induce potentiostat instability causing oscillation of the WE using conventional three-electrode setups. Such instabilities can typically arise from high solution resistances in cell constructions which employ rather narrow or long connecting capillaries between working, counter and reference electrodes. The first obstacle

S. J. Ashton, *Design, Construction and Research Application of a Differential Electrochemical Mass Spectrometer (DEMS)*, Springer Theses,
DOI: 10.1007/978-3-642-30550-4_4, © Springer-Verlag Berlin Heidelberg 2012

therefore, was to diagnose and resolve this crucial issue in order to achieve reliable and precise potential control of the WE, without which we cannot perform experimentation. Once a stable control of the electrochemistry was achieved, the influence of electrochemical cell parameters, namely the electrolyte flow rate through the cell, was investigated. This was of prime interest because the electrolyte flow rate not only influences the electrochemistry at the WE, but also the QMS response impacting the DEMS instrument as a whole. More specifically, the electrolyte flow rate affects the observed diffusion limited current, I_{dl}, the response time and correlation between the faradaic and observed QMS mass ion current, and finally, the DEMS cell collection efficiency. In the following sub-sections, the influence of experimental parameters on the characteristics of the dual-cyclone thin-layer electrochemical cell measurement is presented and assessed. The methodologies employed here to characterise this cell should also be useful for the optimisation, assessment and comparison of different DEMS electrochemical cell designs that may be fabricated in the future.

4.1.1 Potential Control

The ability of the potentiostat to precisely and reliably control the potential of the WE is of vital importance to electrochemical experimentation. The potential control of the WE in the dual-cyclone thin-layer electrochemical flow cell setup fabricated for this DEMS instrument, however, is unstable and the WE potential oscillates using standard three-electrode electrochemical cell arrangements. This essentially means that electrochemical experimentation is impossible under the desired experimental conditions, and a solution to achieve reliable potential control was of vital importance to the operation of the DEMS instrument.

The instability and oscillation of a potentiostat can arise as a result of a number of issues regarding the characteristics of the electrochemical cell configuration, such as: (1) a high RE impedance, (2) high CE resistance (3) high cell capacitance and (4) too small of a solution resistance ca. $<1\ \Omega$ between WE and Luggin-capillary, typically arising from excessive *iR*-compensation of the solution resistance. The most common causes of potentiostat stability are high cell resistances between electrodes, and therefore often the easiest solution to stabilise a potentiostat is to reduce the resistances between the WE, RE and CE by simply positioning the electrodes closer together. Unfortunately, this is largely not possible in the DEMS electrochemical cell, although we can reduce resistances by widening the connecting capillaries of the electrode compartments and/or increasing the concentration of the electrolyte. These modifications certainly help to reduce the cell resistances but not to the extent that reliable potentiostat stability is achieved. In such cases where high resistances are a problem but the cell design cannot be changed, there are certain modifications to the three-electrode electrochemical cell setup, such as those described in Ref. [1], that can be used to stabilise the potentiostat despite the high cell resistances. For example, instability arising from

Table 4.1 DEMS electrochemical flow cell resistances between WE, CE, RE electrode compartments and membrane interface sampling compartment (SC)

Electrode position	Electrolyte resistance/ kΩ		
	$HClO_4$		H_2SO_4
	0.10 mol dm^{-3}	1.00 mol dm^{-3}	0.50 mol dm^{-3}
CE ↔ WE	12	1.4	2.5
RE ↔ WE	17	2.7	4.9
CE ↔ RE	29	5	10
WE ↔ SC	7	–	–

Solution resistances were determined using AC voltammetry (5 mV, 5,000 Hz at 20 mV s^{-1}) in a two-electrode electrochemical cell setup employing platinised Pt wires. The cell resistances are generally rather large but can be reduced by employing more concentrated electrolyte

high RE impedances can be reduced by employing a low impedance reference probe (i.e. Pt wire) positioned close to the WE that is coupled to the RE via a capacitor, or via a high frequency shunt in the form of a capacitor connecting RE and CE [1]. A full explanation of the origins of potentiostat instability (such as the electronic circuitry theory and physics) and its resolution is some way beyond the scope of this chapter, however, a comprehensive overview of the circumstances that cause this instability may be found in references [2, 3]. Unfortunately, these methods were found (by trial and error) not to satisfactorily resolve the issues of the DEMS electrochemical cell, suggesting that despite high cell resistances being less than ideal, they were not necessarily the main cause of the instability.

The first step toward diagnosing the origins of the potentiostat instability of the dual-cyclone thin-layer DEMS cell was to measure the solution resistances between the WE, CE and RE. The solution resistances determined between the electrode compartments of the dual-cyclone thin-layer flow cell in various electrolyte and concentrations are presented table 4.1. These were determined using a two-electrode setup and AC voltammetry.

It is immediately apparent that the electrochemical cell resistances in 0.10 mol $dm^{-3}$are quite large, and can therefore be partly attributed to the potentiostat instability. The resistance between WE and RE, however, may not be quite that problematic in consideration that problems resolved in the literature examples were typically >50 kΩ [1], although the CE to WE resistance in DEMS electrochemical cell is much higher than typical cells that do not isolate the CE from the WE compartment (such as RDE cells). By employing more concentrated and thus more conductive electrolyte, these cell solution resistances are roughly reduced by a factor of ~10, to much more reasonable values. Despite these reductions in cell resistances, potentiostat stability is still not achieved using a standard three-electrode setup and therefore an alternative modification to the electrochemical half-cell configuration was required.

The origins of the potentiostat instability was in fact found to arise because of too small of a solution resistance, R_s, ≤ ca. 1 Ω between the WE and Luggin-capillary. Potentiostats typically require a certain amount of R_s in order that

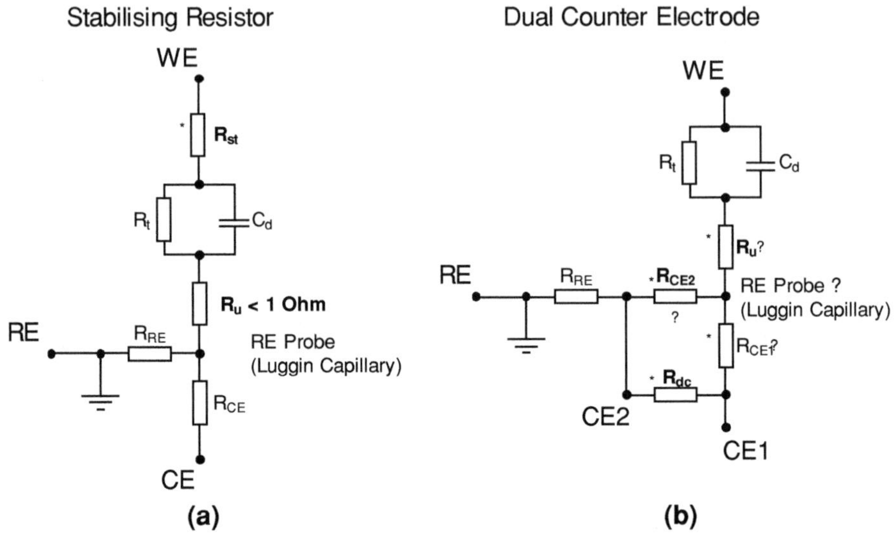

Fig. 4.1 Equivalent circuits for the DEMS electrochemical cells that employ: **a** the stabilising resistor, or **b** the dual counter electrode setups to stabilise the potential control of the WE. Modifications to standard equivalent circuit are marked with a *. Key: stabilising resistor (R_{st}), charge transfer resistance (R_t), dual counter resistor (R_{dc}) and uncompensated (R_u), reference electrode (R_{RE}) and counter electrode (R_{CE}) solution resistances, and double layer capacitance (C_{dl})

conditions of potentiostat oscillation are not met, which is why 100 % iR-compensation is generally not possible (using analogue iR-compensation schemes) although it can be minimised to give errors of ~ 1 mV (the typical error of a RE) by choosing certain cell parameters [2]. The DEMS electrochemical cell, however, does not possess a Luggin-capillary in the traditional sense of a narrow capillary leading to the WE. Instead, the electrochemical flow cell lines between the WE and RE are analogous to the Luggin-capillary, and the point at which these cell lines intersect the WE and CE current path can be considered to be the Luggin-capillary entrance. This would lie somewhere between the electrochemical flow cell inlet entrance to the WE compartment and the holes joining the WE compartment to membrane interface, although this cannot be defined precisely. Because this distance is relatively little <0.2 mm the solution resistance is too small and the conditions at which the potentiostat oscillates is met, which poses somewhat of a dilemma in DEMS because the thin-layer is crucial to the performance of the instrument i.e. we cannot simply increase the layer thickness to the extent that the solution resistance becomes significant enough without impeding on the DEMS instruments performance. An alternative solution was therefore needed.

The solution in the case of this DEMS electrochemical cell setup was to simply stabilise the potentiostat by artificially increasing the solution resistance by adding a stabilising resistor between the WE and potentiostat, the equivalent circuit of which is shown in Fig. 4.1a.

The size of resistor required to stabilise the potentiostat is most easily found by trial and error, however, it is advantageous to use the lowest reliable resistance value possible in order to minimise the uncompensated WE resistance. For reference, the potentiostat was stabilised using a resistance of ~110, ~25 and ~25 Ω in 0.1 and 1.0 mol dm^{-3} $HClO_4$, and 0.05 mol dm^{-3} H_2SO_4, respectively. In general, these errors are unacceptably large, introducing WE potential errors of 110 and 25 mV at 1 mA of current (without employing *iR*-compensation) which is particularly undesirable in cyclic voltammetry when electrode currents are potential scan rate dependent. Of course *iR*-compensation could be applied using the potentiostat, however, compensation using conventional analogue potentiostats that employ positive feedback compensation schemes may not be beneficial at all due to the origin of the potentiostat instability of this cell i.e. by trying to compensate for the stabilising resistor we quickly reintroduce potentiostat oscillation and lose potential control of the WE. It is worth noting that these problems may be avoidable using a potentiostat that employs a digital *iR*-compensation scheme.

An alternative *iR*-compensation technique that does not intrude upon the potentiostat control amplifier circuitry is the negative impedance convertor device (NID) [4], the technical details (i.e. circuitry) of which were presented previously in Chap. 3 Sect. 3.6.1.1. The NID device was successful in reducing the uncompensated stabilising resistance, R_{st}, from ~110 to ~60 Ω in 0.10 mol dm^{-3} $HClO_4$ and from ~25 Ω to an acceptable ~10 Ω in 1.0 mol dm^{-3} $HClO_4$ and 0.5 mol dm^{-3} H_2SO_4 electrolytes. This more reasonable uncompensated resistance can then be corrected post measurement to obtain close to the actual potential current relationship. The combination of the stabilising resistor and NID to apply *iR*-compensation allows potentiostat stability along with the measurement (using AC voltammetry) and minimisation of the uncompensated solution resistance. Potentiostat instability, however, is not specific to the dual-thin layer electrochemical cell fabricated for this DEMS instrument and other methods have been used to stabilise the potentiostat, such as the dual-counter electrode cell setup.

In the dual-CE cell arrangement, a second CE is placed on the DEMS cell outlet and is coupled to the primary CE (on the inlet) via a resistor, as shown in the equivalent circuit in Fig. 4.1b. Although it is not entirely clear how exactly this cell configuration functions, it presumably produces a similar effect to the stabilising resistor, of increasing the effective solution resistance. The second CE seemingly shifts the effective position of the Luggin-capillary entrance so that the solution resistance is increased to the WE by allowing a proportion of the cell current to pass through the cell outlet to the second CE. The location where the RE probe intercepts the current path between the WE and CE is shifted closer to the RE, increasing the solution resistance and stabilising the potentiostat. The dual-CE setup, however, cannot be considered to be a well-established or even documented solution for resolving potentiostat stability (i.e. the author could not find any literature outside DEMS on the dual-CE configuration) and it is not clear how exactly this cell configuration is working, or whether, for example, the solution resistance is current dependant, changing the position of the Luggin-capillary and therefore uncompensated solution resistance. Furthermore, there have been no indications in previous DEMS literature of the typical uncompensated solution resistance or whether it can be measured at all in such a configuration and if so, whether *iR*-compensation of the solution resistance is possible. It is therefore proposed that the stabilisation of a potentiostat used in the DEMS electrochemical dual thin-layer flow cell via a stabilising resistor on the WE connection to the potentiostat in combination with *iR*-compensation using the NID (or possibly via a potentiostat that employs digital *iR*-compensation scheme), is a more transparent, reliable and understood solution to resolve potentiostat instability compared to the dual-CE arrangement employed in previous studies.

To summarise, the origins of the potentiostat stability in the three-electrode setup of the dual-cyclone thin-layer cell was found to largely arise from a small solution resistance between the WE and the effective Luggin-capillary rather than large cell resistances. By employing a stabilising resistor between the WE and potentiostat, the solution resistance of the electrochemical cell can be effectively increased, therefore stabilising the potentiostat allowing potential control of the WE. By using a NID for *iR*-compensation, the uncompensated resistance introduced by the stabilising resistor can be reduced to acceptable values of $\sim 10\ \Omega$. Although in DEMS we are indeed very interested in the in situ observation of electrochemical reaction species using mass spectrometry, this should not be at the expense of the quality and precision of the electrochemical measurement.

4.1.2 Effect of Electrolyte Flow Rate

The following sections demonstrate the methods of characterising the influence of the electrolyte flow rate on the: diffusion limited current; response time and correlation of the QMS mass ion currents; and the collection efficiency of the dual thin-layer flow cell.

4.1.2.1 Diffusion Limited Current

The first and probably most straightforward characteristic of the dual-cyclone thin-layer electrochemical flow cell to determine, is the relationship between the diffusion limited current for a reaction process, I_{dl}, and the volumetric electrolyte flow rate, V_f. Diffusion limited currents arise when the rate of electrochemical reaction is limited by the diffusion of reactants from the bulk to the electrode surface, rather than the kinetics of the electrode reaction. In hydrodynamic cells such as the DEMS flow cell, the I_{dl} is related by a certain expression to the V_f of the electrolyte containing the reaction species over the WE surface. In general, such hydrodynamic cells may be separated into two categories: (1) uniformly accessible electrodes, such as the RDE and (2) non-uniformly accessible electrodes, such as the channel flow cell. The principle difference between these two categories is whether or not the I_{dl} scales proportionally to the geometric area of

Fig. 4.2 An example CV series for ORR on polycrystalline Pt at 20 mV s^{-1} for various electrolyte flow rates in the dual-cyclone thin-layer flow cell. The diffusion limited current is reached at ~0.25 V_{RHE} and increases with increasing volumetric flow rates

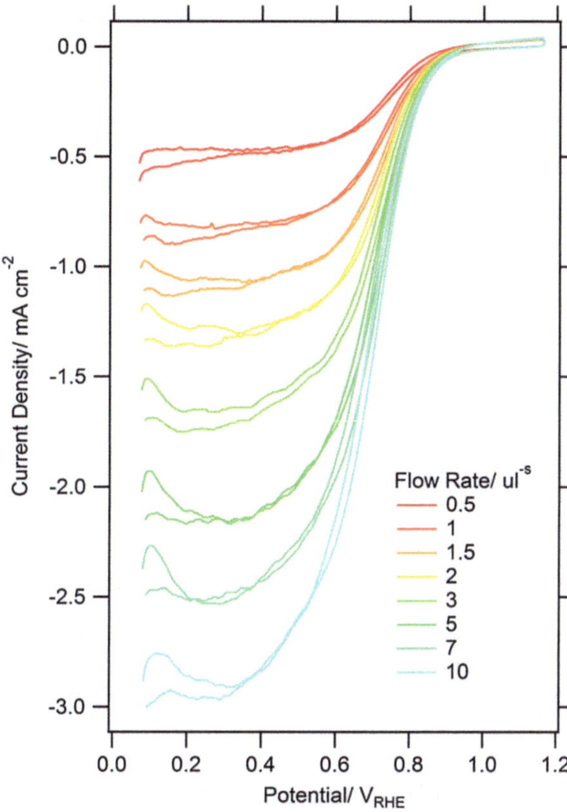

the electrode: if the I_{dl} scales then the electrode is considered uniformly accessible. The dual cyclone (and linear) thin-layer flow cell(s) fall into the latter non-uniformly accessible electrode category. In the case of the dual-cyclone thin-layer flow cell, which does not possess a typical flow regime over the WE, it is particularly important to both understand the relationship between the diffusion limited current and electrolyte flow rate in order to compare the dual-cyclone cell to more common flow regimes. This is especially true for experiments where the reactant of interest is of low concentration in the electrolyte.

The relationship between the I_{dl} and the V_f. can be most easily determined by measuring the CV for the oxygen reduction reaction (ORR) of an oxygen saturated electrolyte at a series of volumetric electrolyte flow rates. An example series of CVs for the (ORR) on polycrystalline Pt electrode in the dual-cyclone thin-layer flow cell is presented in Fig. 4.2. The CVs clearly exhibit the same characteristic features observed for the ORR using a RDE i.e. a kinetically limited region at potentials >0.8 V_{RHE}, a diffusion limited region below 0.4 V_{RHE} and the so-called mixed region lying between.

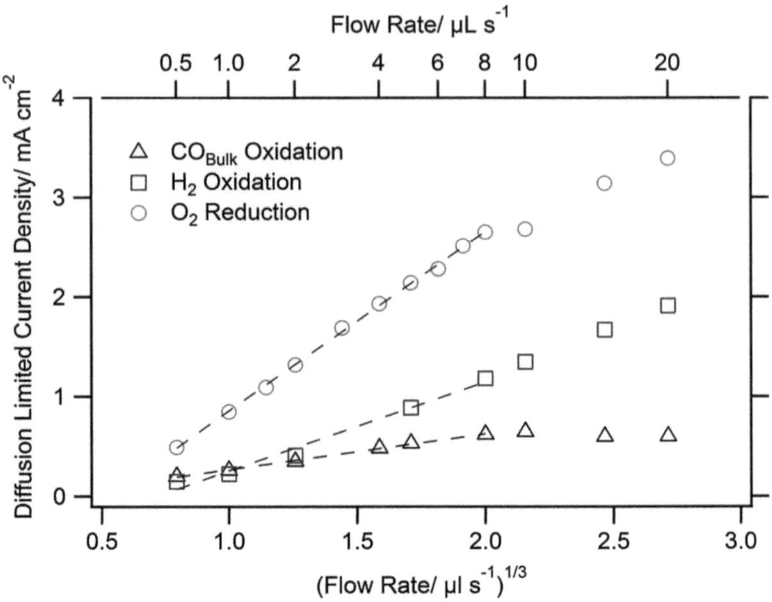

Fig. 4.3 The dependence of the diffusion limited current, I_{dl}, on the electrolyte flow rate for the ORR, HOR and CO bulk oxidation reaction in the dual-cyclone thin-layer flow cell. The diffusion limited current scales with the volumetric flow rate according to a power of 1/3 for flow rates below 10 µl s^{-1}

The maximum diffusion limited current occurs at ~0.3 V_{RHE} for each flow rate similarly to that observed using RDE (when increasing rotation), and increases with increasing electrolyte flow rate. For non-uniformly accessible hydrodynamic electrodes the I_{dl} scales according to a number of constants multiplied by the volumetric flow rate raised to a power. For example, the relationship between the diffusion limited current and the volumetric flow rate for a channel electrode is given by the following equation Ref. [5]:

$$I_{dl} = 0.952nFc_{\infty}D^{2/3}(h^2d)^{-1/3}wx^{2/3}V_f^{1/3}$$

where, n = no. of electrons, F = faraday constant, D = diffusion coefficient for the reacting molecule, h = 1/2 height of flow channel, d = flow channel diameter, w = WE width and x = position relative to start of electrode and V_f = volumetric flow rate (flow parameter). These terms will be constant for a certain reaction in the cell, and therefore the I_{dl} will essentially scale with the V_f. raised to a certain power. Note that this is generally true for all non-uniformly accessible electrodes [5]. For the dual-cyclone thin-layer flow cell the I_{dl} was found to scale with V_f. to the power of a 1/3, as shown in Fig. 4.3.

The I_{dl} scales linearly with the volumetric flow rate, within the range between 0.5 and 10 µl s^{-1} before deviating at higher flow rates. This value of 1/3 is comparable to the value of 0.4 given in Ref. [6] for another dual-cyclone thin-layer cell (although with presumably slightly different dimensions) which was suggested to be analogous to the wall-jet flow cell. The wall-jet electrode, however, scales with the volumetric flow rate according to the power of 3/4 [5]. The exact values, however, will be dependent on the cell dimensions. Finally, the loss of linearity observed in Fig. 4.3 at flow rates ≥10 μl s^{-1} suggests that the flow of electrolyte becomes less stable (perhaps indicating more turbulent flow) and that lower flow rates should therefore be favoured for DEMS experimentation.

4.1.2.2 Response Time

The QMS response time, or rather the delay between the formation of a species at the WE and its subsequent detection by the QMS was measured as a function of the electrolyte flow rate. This delayed response is largely determined by the finite time required for the reaction species to be transported away from the WE to the membrane surface, and is therefore strongly dependent upon the electrolyte flow rate. The relationship between the volumetric flow rate and the QMS response time is first shown quantitatively in Fig. 4.4 via the linear sweep voltammogram (LSV) of CO-bulk oxidation and corresponding mass ion current, $I_{m/z\ =\ 44}$ on polycrystalline Pt at various electrolyte flow rates.

Here we see that the time difference between the generation of carbon dioxide and the mass ion current response predictably decreases with increasing flow rate, owing to the reduced transport time of reaction products in solution from the WE to the membrane. At the very low flow rate of ~0.5 μl s^{-1} the QMS response time is some 15 s long, whereas this is reduced to below 1 s at 10 μl s^{-1}. In order for the QMS measurement signal to correlate with the electrode potential, the time of the mass ion current must be corrected for in post measurement data treatment. It also evident from Fig. 4.4, that the reproduction of the features observed in the electrochemistry by the QMS current is also dependent on the flow rate (this will be addressed via the correlation in the next section). The precise dependence of the QMS response time on the electrolyte flow rate through the dual-cyclone thin-layer cell was determined using potential step chronoamperometry of bulk CO oxidation on polycrystalline Pt, and is presented quantitatively in Fig. 4.5.

The QMS response time decays in a power law fashion from around 15 s to below 1 s as the electrolyte flow rate increases >10 μl s^{-1}. The large error bars present at flow rates below 4 μl s^{-1} demonstrates that the QMS response time is very sensitive to slight fluctuations in the precise flow rate in this range. These can, for instance, arise from minor disturbances of the DEMS flow cell PTFE tubing which induces turbulence in the electrolyte within the thin-layers. At flow rates above 4 μl s^{-1} the QMS response time gradually approaches a steady value below 1 s and the response time is significantly more reproducible. The correction of the QMS data in potential and time therefore, can be acceptably performed without error. If one was to only consider the QMS response time in choosing the optimum electrolyte flow rate, values greater than 4 μl s^{-1} would be acceptable. There are

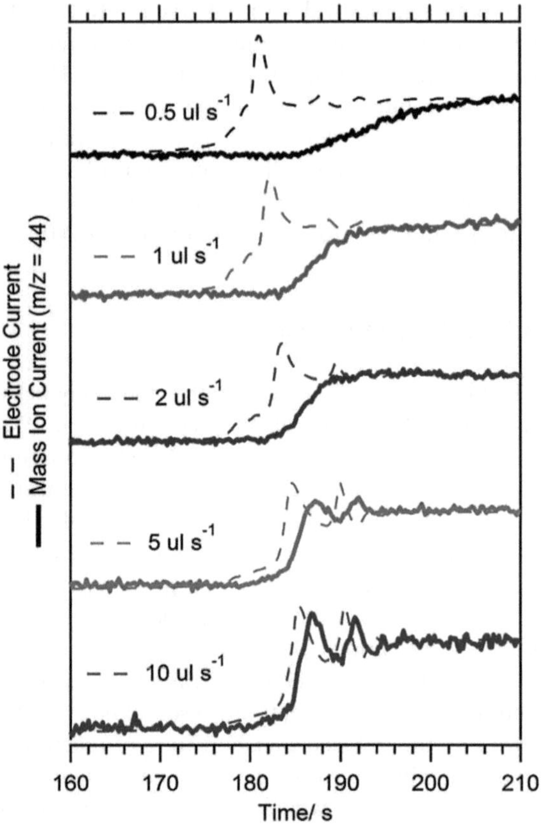

Fig. 4.4 A graph qualitatively highlighting the difference between the faradaic electrode current and the corresponding $I_{m/z\,=\,44}$ response during a CO-bulk oxidation measurement on polycrystalline Pt for the dual-cyclone thin-layer flow cell performed at various electrolyte flow rates. The QMS response time decreases whilst the reproduction of the electrochemical features improves with increasing electrolyte flow rate

however, other characteristics of the DEMS instrument which must also be considered, such as correlation and collection efficiency.

4.1.2.3 Correlation

The degree of correlation between the electrode current arising from a certain reaction at the WE, and the subsequent QMS response to the electro-generated species is a strong function of both the electrolyte flow rate and potential scan rate. The difference between the CV for CO bulk oxidation on Pt and the mass spectrometric cyclic voltammogram (MSCV) for $I_{m/z\,=\,44}$ is first illustrated qualitatively at various electrolyte flow rates in Fig. 4.6. The correlation between the CV and MSCV improves as the electrolyte flow rate through the cell is increased. This is indicated for instance, by the reproduction of the CO-bulk oxidation peak observed in all the CVs at $\sim$1.0 V_{RHE} that is only resolved in the MSCV at electrolyte flow rates exceeding 10 μl s^{-1}. Furthermore, the presence of carbon dioxide remaining in the MSCV (and therefore the electrochemical cell) during the

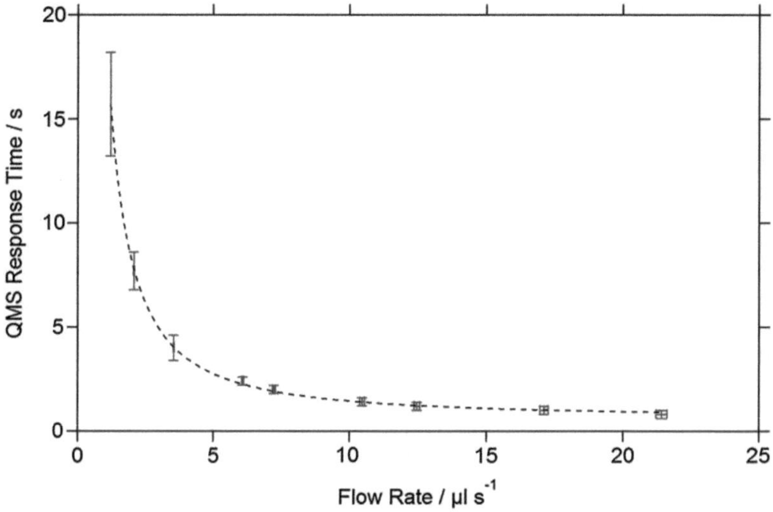

Fig. 4.5 The dependence of the QMS response time for bulk CO-oxidation on the electrolyte flow rate through the dual-cyclone thin-layer flow cell. The large error bars indicating the variability in the response time suggests slight but significant fluctuation in the flow of electrolyte through the cell at low flow rates

reverse potential sweep after CO oxidation has ceased at ~0.9 V_{RHE} is longer at lower electrolyte flow rates. This effect is associated with the residence time of the reaction product in the electrochemical cell compartment, which at lower electrolyte flow rates is more prone to diffusion and mixing in the electrochemical cell thin-layers. Furthermore, the degree of correlation will also be dependent on the potential scan rate of the CV, whereby fast scan rates or electrode processes are inherently more difficult to reproduce. Such qualitative observations, however, are somewhat unsatisfactory by themselves and it would be far more useful if the correlation between the CV and MSCV is expressed in numbers so that optimum experimental parameters can be precisely identified. Fortunately, there are a number of statistical methods that exist, such as *correlation coefficient* or *goodness of fit*, which are designed to quantify the degree of fitting, or correlation between two data sets. Typically, these statistical methods are used to assess the fit of a mathematical model to true experimental data. By defining the measured WE current as our true data and the QMS response as our modelled data, the correlation of the QMS response can be assessed quantitatively.

The dependence of the goodness of fit for the MSCV (m/z = 32) on the electrolyte flow rate at various potential scan rates during the ORR on polycrystalline Pt is give in Fig. 4.7. These quantitative observations indicate that the correlation of the MSCV improves with increasing electrolyte flow rate, and decreasing potential scan rate. In general, the correlation of the MSCV was relatively poor for the high CV scan rates of 50 and 100 mV s^{-1} at all electrolyte flow rates.

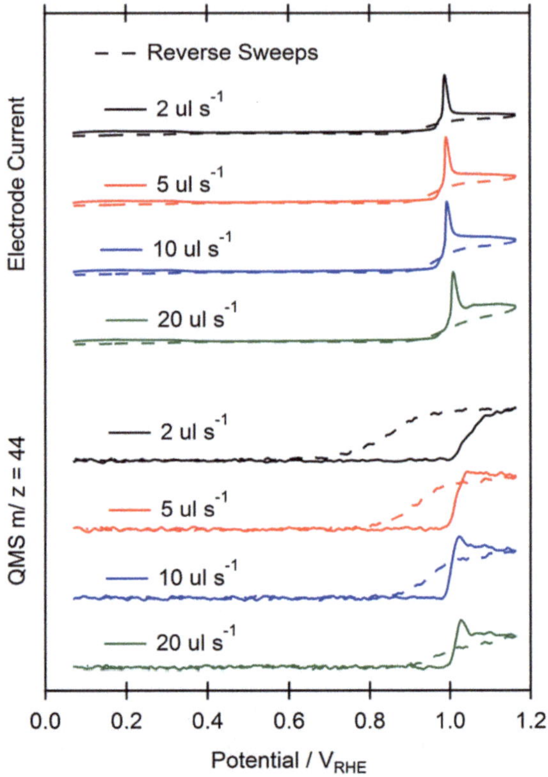

Fig. 4.6 CV and MSCV of CO_2 (m/z = 44) during CO-bulk oxidation at various different electrolyte flow rates for the dual-cyclone thin-layer flow cell. The correlation of the QMS mass ion current with the electrode current improves with increasing electrolyte flow rate. The residence time of the electrochemical reaction species in the electrochemical cell also decreases

Acceptable correlation of the MSCV can be, however, achieved at scan rates of 5, 10 and 20 mV s^{-1} by employing flow rates of at least 3 μl s^{-1}. The exact values of the correlation will of course be dependent on the nature of the electrochemical reaction and the features of the CV, of which the ORR is a rather simple example. In order to identify an optimum electrolyte flow rate, we must finally assess the DEMS cell collection efficiency.

4.1.2.4 Collection Efficiency

The DEMS cell collection efficiency is simply the fraction of reaction products produced at the WE that then evaporate through the PTFE membrane interface into the vacuum system to be detected by the QMS. The collection efficiency, therefore, has a direct influence on the detection limit and sensitive of the QMS and is ideally close to 1.0. Although high collection efficiencies >0.9 can be obtained with electrolyte convection, any flow of electrolyte through the electrochemical cell will reduce the retention time of the electrochemical species above the membrane interface. The time frame in which the electrochemical species can

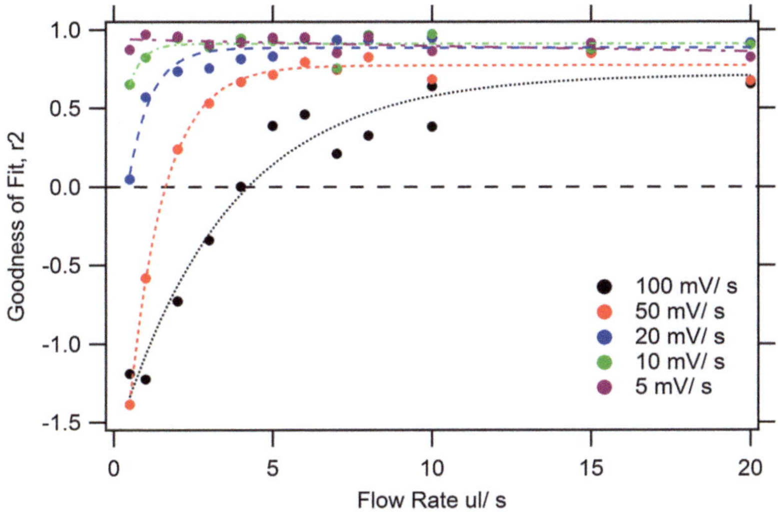

Fig. 4.7 A plot demonstrating the dependence of the goodness of fit of the MSCV to the CV of the ORR, on the electrolyte flow rate at various potential scan rates for the dual-cyclone thin-layer flow cell. Acceptable correlation (close to 1.0) between the electrode current and QMS mass ion current is achieved at potential scan rates $\leq$20 mV s^{-1} and at flow rates $\geq$4 μl s^{-1}

therefore diffuse to the membrane surface is diminished and thus the collection efficiency falls. The DEMS cell collection efficiency obtained from CO bulk oxidation on polycrystalline Pt is presented as a function of electrolyte flow and potential scan rate is shown in Fig. 4.8. Note that the method of determining the DEMS cell collection efficiency, *N*, required for calibration of the instrument is outlined later in Sect. 4.4.

The DEMS cell collection efficiency decreases as the electrolyte flow rate through the cell is increased. A collection efficiency of ~0.9 can be achieved using with very low electrolyte flow rates, however, this decreases rapidly until 4 μl s^{-1} before steadying and plateau-ing at flow rates >15 μl s^{-1}. Furthermore, the collection efficiency is independent of potential scan rates of 50 mV s^{-1} or lower, and flow rates greater than 3 μl s^{-1}. Note that the discrepancies at high scan rates and low flow rates arise due to the failure for the total QMS peak response to be produced in the corresponding CV, and the integration of the QMS data is therefore incomplete. The collection efficiency is therefore concluded to be independent of the potential scan rate. On one hand, a high collection efficiency is desired to maximise the instrument sensitivity, which would favour electrolyte flow rates <3 μl s^{-1}, however, the sensitive dependence of the collection efficiency on the exact electrolyte flow rate <3 μl s^{-1} may lead to difficulties in terms of reproducibility for quantitative analysis if the flow rate is not absolutely stable. Consequently, an electrolyte flow rate <3 μl s^{-1} is not considered to be optimum. Finally, the performance in terms of collection efficiency at a flow rate of 5 μl s^{-1} is comparable to a value of 0.3 provided in Ref. [6].

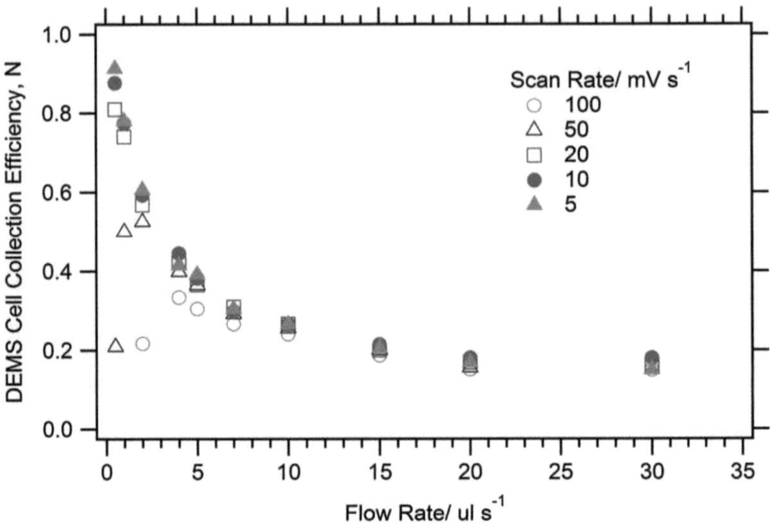

Fig. 4.8 A plot illustrating the dependence of DEMS cell CO_2 collection efficiency, *N*, on electrolyte flow rate at various potential scan rates for the dual-cyclone thin-layer flow cell. The DEMS collection efficiency decreases with increasing electrolyte flow rate, beginning to stabilise at rates ≥ 4 μl s^{-1} and is independent of potential scan rate

4.2 Performance of the Membrane Interface Material

There are two properties of the PTFE membrane that are critical to the performance of the DEMS instrument: (1) the total gas flux, and (2) the selective permeability. The total flux of gas through the membrane from the aqueous phase to the vacuum determines the size of the vacuum system necessary and/or the geometric area of the membrane interface as discussed in Chap. 3. The selective permeability of the membrane meanwhile, determines if and how easily certain species are able to evaporate through into the vacuum system. Presented in this section are methods in which to evaluate both of these membrane characteristics.

There are a number of methods in which the total gas flux through a membrane may be determined to evaluate if the membrane is appropriate. For example, if the pumping speed of a simple vacuum system is well-known, then the gas flux can be easily determined by simply measuring the vacuum system pressure. There are, however, two problems associated with this method. The first is that the total gas flux through the membrane will be to some extent dependent on the vacuum system pressure, and the secondly there is the risk of destroying the turbo pumps if the membrane is inappropriate. A rather simple and robust test setup to assess the membrane performance is therefore somewhat desirable. The total gas flux through the membrane material may be evaluated using a straightforward setup consisting of a small vacuum chamber, pressure gauge, roughening pump and the DEMS electrochemical cell membrane interface, as illustrated in Fig. 4.9.

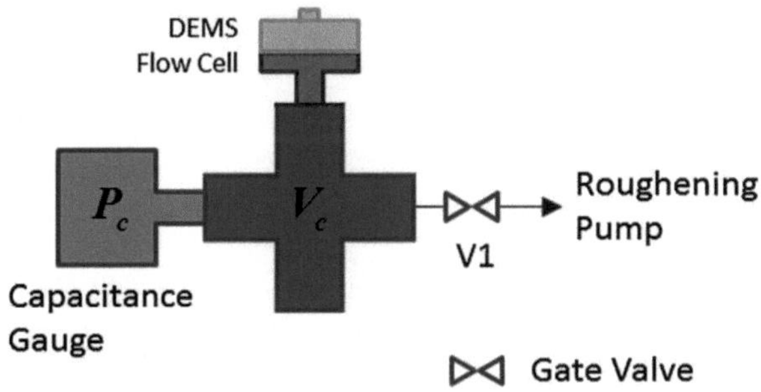

Fig. 4.9 Membrane interface material vacuum test setup.The roughening pump is used to evacuate the vacuum chamber volume, whilst the pressure is monitored using a capacitance gauge. By shutting the gate valve V1 the total gas flux through the membrane is measured as a function of the vacuum pressure

The membrane sample is setup in the DEMS flow cell to form the membrane vacuum interface, and electrolyte is passed through the cell (almost as if a DEMS experiment were to be performed). The operating roughening pump may then be allowed to evacuate the volume by opening gate valve (V1) which will have one of two consequences. Either a vacuum is not achieved, in which case the membrane material is completely unsuitable, or a vacuum is achieved to give a vacuum pressure <50 mbar, in which case the membrane may be suitable.

If the pumping is then rapidly closed off using the gate valve (V1), the corresponding increase in the pressure of the vacuum volume can be used to determine the flux of gas through the membrane according to the following equation:

$$J = V_c \frac{\delta P_c}{\delta t}.$$

In order to extract the total gas flux, the vacuum pressure must be recorded accurately as a function of time. This is achieved using the QMS calibration and measurement setup described previously in Chap. 3. The total gas flux, J, through a number of the microporous PTFE membrane candidates is plotted as a function of the vacuum pressure in Fig. 4.10.

This plot shows that the gas flux through the membrane decreases with increasing vacuum pressure. This trend approximately follows an exponential decay, and through fitting the total gas flux may be extrapolated to vacuum pressures expected for the vacuum system ~0.01 mbar. Through fitting a number of these measurements, and extrapolating to a vacuum pressure 0.01 mbar, gas fluxes approximating those of the membrane when attached to the vacuum system can be obtained. These values are given in Table 4.2.

The total gas flux extrapolated from the measurement data to the DEMS vacuum system pressure for the Gore and Donaldson membrane pore sizes of ≤0.4 μm are comparable to the calculated values of 0.09 mbar l s^{-1} at 25° C [7] and 0.4 mbar l s^{-1} at 40°C [6]. There is no particular trend concerning the

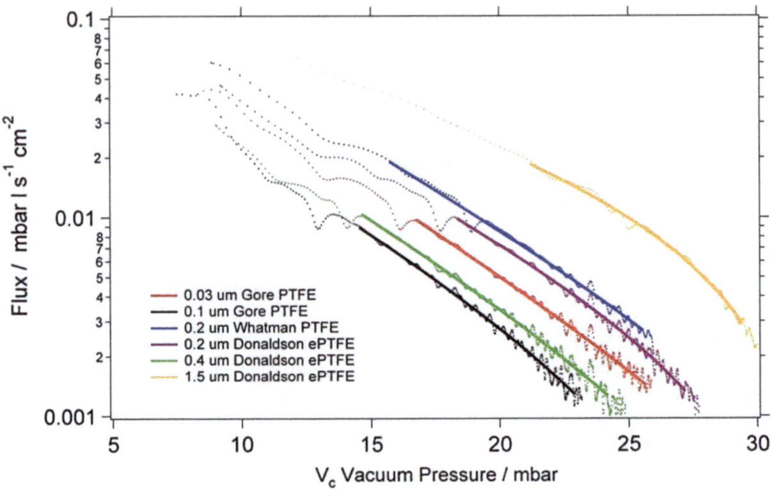

Fig. 4.10 Total gas flux through porous membrane into an evacuated volume plotted as a function of the evacuated volume pressure. The flux of gas permeating the microporous membrane is dependent on the vacuum pressure applied

Table 4.2 The total gas flux values extrapolated to 0.01 mbar vacuum pressure at 5 μl s^{-1} electrolyte flow rate

Manufacturer	Pore size/μm	Thickness/μm	Total gas flux at 0.01 mbar vacuum pressure
Gore	0.03	127 ± 9	0.1146
	0.1	124 ± 11	0.1463
Whatman	0.2		0.3866
Donaldson	0.2	89	0.2351
	0.4	152	0.1211
	1.5	152	0.9328

The total gas fluxes of the potentially suitable membrane interface materials lie between 0.1 and 1.0 mbar l s^{-1} cm^{-2}

relationship between the micropore size and total gas flux in this data, however, this is likely attributed to the precision of the measurement. In practical terms there is little difference between pore sizes for the microporous Gore membranes in DEMS.

The second property of the membrane of relevance to DEMS is the selectivity. The selectivity of the membrane will not only determine which species evaporate into the vacuum system, but also how easily, which in turn directly influences the sensitivity and detect-ability of the DEMS instrument. A measure of the ability of a certain species to pass through the membrane is the DEMS collection efficiency. A method for determining the DEMS cell collection efficiency, *N*, for CO_2 using CO-Stripping is later described in Sect. 4.4. There is, however an alternative method whereby the electrolyte containing a certain concentration of is simply flowed through the cell (no electrochemistry is involved), and the relevant mass

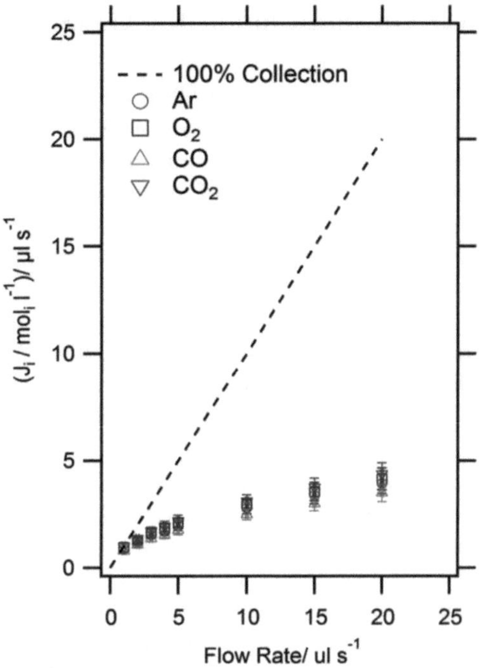

Fig. 4.11 Mass ion current converted to flux normalised to concentration of species in aqueous solution. The permeability of the dissolved gaseous species are similar within experimental error, and are not limited by the membrane permeability but rather the diffusion of species from the bulk electrolyte to the membrane pore entrances

ion current is monitored as a function of flow rate. The membrane sample compartment collection efficiency, *f*, is thus given by:

$$f = \frac{I_{m/z=i}}{K_i^0 c_i v_f}.$$

where c_i = concentration of species in solution and v_f = volumetric flow rate through the cell. This experiment was performed for a number dissolved gaseous species, and the collection efficiency of the membrane sampling compartment converted to flux is plotted as a function of flow rate in Fig. 4.11.

The fluxes of gaseous and volatile species are similar within the experimental uncertainty. For very volatile species the flux through the membrane is limited by the diffusion of gaseous species to the membrane pores, and consequently it is unlikely that any significant difference between different membrane materials to the detection of gases can be observed. The membrane interface collection efficiency therefore is as much, if not more of a measure of the performance of the membrane interface design rather than the membrane material. It would, however, be worthwhile to determine and compare the membrane interface collection efficiency for less volatile species which may be limited by their passage through the membrane should the experiment desire so. For less volatile species such organic molecules, the permeability through the membrane is significantly less, which is nicely illustrated and discussed in Ref. [6].

4.3 Optimisation of the QMS

The optimisation of the QMS parameters is necessary in order to maximise the DEMS instrument sensitivity and stability, producing better quality QMS data with optimal signal-to-noise ratio and therefore higher detection and quantification limits. Whilst some brief guidelines for optimising the QMS are presented in the operating manuals [8–10], a more detailed and qualitative description with figures that clearly demonstrate the effects of the QMS tuning parameters are presented here.

The optimisation of the ion source begins by admitting a controlled gas leak into the vacuum system. Ideally, this is a molecule of interest, such as carbon dioxide, so that vacuum conditions similar to those during the DEMS instrument operation are obtained. This can be achieved by attaching a gas line (or by simply admitting air) to the vacuum system inlet valve, and partially opening the inlet leak-valve until the appropriate vacuum system pressure is achieved. The QMS ion source settings are then optimised using the *Quadstar 32-bit Tune Up* application in the *Tune Ion Source* window. Here, the QMS settings can be manipulated and the effect on the observed mass ion current can be evaluated in real time. QMS parameters can then be systematically varied and the averaged mass ion current response value recorded. In order to tune the QMS sensitivity, it is crucial to first optimise the ion source parameters followed by the quadrupole and SEM.

4.3.1 Ion Source Parameters

The ion source settings can have a significant impact on the sensitivity of the QMS because the ion source settings will determine how many ions are produced, and are guided to the quadrupole for mass selection. This can involve modifying the ion acceleration voltage, however, the emission current, along with the extraction and focus plate voltages are the more common parameters required to be optimised.

4.3.1.1 Emission Current

The emission (or electron) current determines how many electrons are emitted from the ion source filament(s). Typical emission currents for the cross-beam (CB) ion source are either exclusively 0.1 or 1.0 mA (see Ref. [10]). The number of ions produced at the ion source is given by the following relationship:

$$I_{k+} = i_- * i_e * s * p_k$$

where I_{k+} = ion current, i_- = emission current, I_e = mean path length of the electrons, s = differential ionization effect cross-section, and p_k = partial pressure of the gas component. In general, an ion current of 1.0 mA is preferred because it

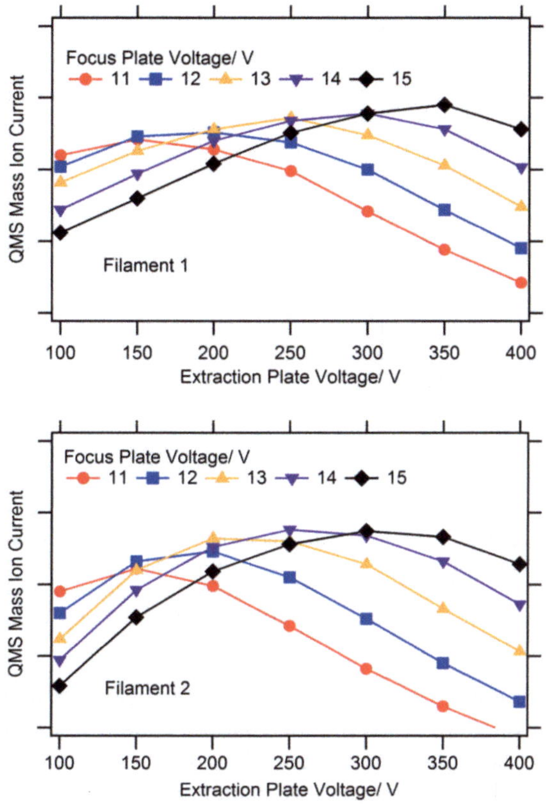

Fig. 4.12 The effect of extraction plate voltage on QMS ion current at various focus plate voltages. A mutual voltage combination for the mass ion current peak maximum should be evaluated based on these curves. Optimum settings based on this data would suggest a focus plate voltage of 14 V and extraction plate voltage of 250 V, although there is no definitively optimal setting for both filaments

will generate ten times more ion current, although a lower emission current may be preferred under certain circumstances, such as when the CB-ions source is operated at pressures approaching 10^{-4} mbar [10].

4.3.1.2 Focus and Extraction Plate Voltages

The focus and extraction plate voltages determine the acceleration and focusing of the ions into the quadrupole mass filter. The focus and extraction plate voltages are not mutually exclusive and so their values can only be tuned by systematically varying the voltages and recording the currents until the optimum parameters are found. The dependence of the mass ion current on the extraction plate voltage at various focus plate voltages for each of the two CB-ion source filaments are shown in Fig. 4.12.

Typical focus plate voltages lie in a narrow voltage range between 11 and 15 V, however, the extraction plate voltage range can be much wider, spanning from 100 to 400 V. From these plots, a mass ion current peak maximum that is common for both filaments at a certain focus and extraction plate voltage should be evaluated

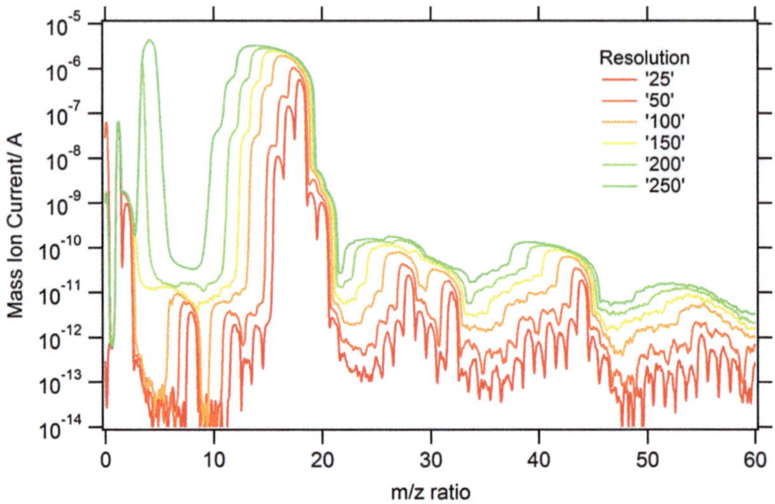

Fig. 4.13 The mass scan analogue spectrum of vacuum chamber background at various QMS resolution values. The selectivity of the quadrupole to a certain mass decreases when increasing the resolution parameter. Higher resolution values are desirable to increase QMS sensitivity when the mass of interest is not surrounded by mass ions with currents of similar magnitude

for normal operation when both filaments are simultaneously running. If more than a single peak maximum exists then the lower focus plate voltage combinations are to be preferred. If the values for each filament are significantly different then mechanical damage, or contamination of the ion optics or filaments may have occurred [8].

4.3.2 Quadrupole and SEM Parameters

Once the ion source settings are optimised, the next step is to optimise the resolution, SEM voltage and deflection plate voltage.

4.3.2.1 Resolution

In analytical chemistry the term resolution usually refers to the degree of peak separation, whereby larger values correspond to better separation; however, in the Quadstar 32-bit software a larger resolution is in fact less separation of the mass ion current peaks (which can cause some confusion), as depicted in the background mass spectrum of the DEMS vacuum system in Fig. 4.13.

The resolution in this case reflects the precision of the mass selection and therefore the number of ions that are transmitted through the mass filter. When

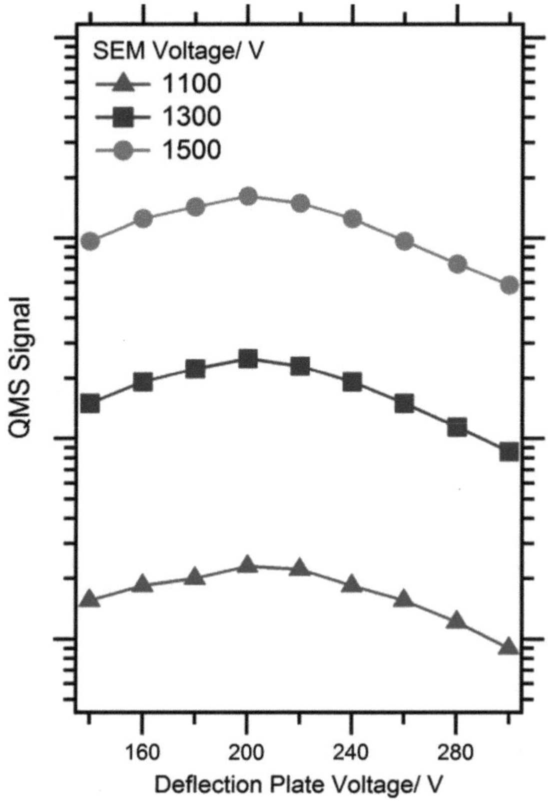

Fig. 4.14 The effect of deflection plate voltage on the QMS mass ion current as a function of SEM voltage. The optimum deflection voltage is only slightly dependant on the SEM voltage. It is therefore not strictly necessary to adjust the deflection plate voltage each time the SEM voltage is changed

operating the QMS in mass ion current mode, the resolution can be raised in order to increase the quadrupole transmission and hence mass ion current and sensitivity of the QMS. In cases where the mass ion signal of surrounding masses are one or two orders of magnitude less than the mass of interest, then the resolution can be safely increased with little interference. The resolution therefore should be optimised for each mass ion current of interest.

4.3.2.2 Deflection Plate Voltage

The deflection plate voltage is responsible for deflecting the mass selected ions exiting the quadruple into the SEM. According to the Pfeiffer vacuum manual, the deflection plate voltage should be optimised each time the SEM voltage is changed [8, 10]. A plot of deflection plate voltage versus mass ion current at various SEM voltages of the QMS used in this DEMS instrument is shown in Fig. 4.14.

In this plot, however, we observe that the peak deflection voltage is ~200 V for a wide range of SEM voltages (1,100–1,500 V). The influence of the SEM voltage on the optimum deflection plate voltage, therefore, does not appear to have a hugely significant impact.

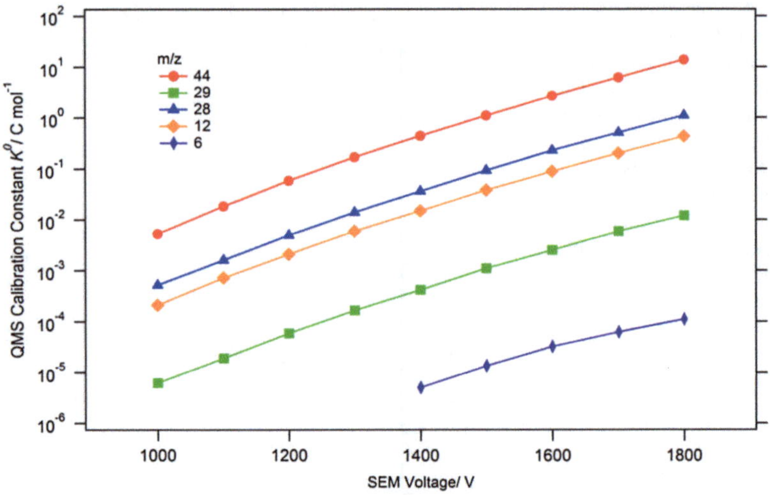

Fig. 4.15 The effect of SEM Voltage on QMS calibration constants for a number of mass ion fragments of CO_2. A 200 V increase in the SEM voltage increases the amplification of the mass ion current signal (and therefore the calibration constant) by a factor of ~10

4.3.2.3 SEM Voltage

The SEM is a device which multiplies the particle beam. The voltage applied to the SEM affects the number of secondary electrons that are produced and thus has a significant effect on the overall amplification of the mass ion current. The effect of the SEM voltage on the measured QMS calibration constant for carbon dioxide mass ion fragments is shown in Fig. 4.15.

Here we see that an increase in the SEM voltage increases the mass ion fragment calibration constant significantly. As a general guide, an increase in SEM voltage of 200 V raises the calibration constant by a factor of ~10. It is, however, inadvisable to operate the SEM at voltages below 1,100 V or greater than 1,600 V (above which the amplification begins to be non-linear). Furthermore, the mass ion current output should in general not exceed 10^{-6} A for any considerable length of time to maintain SEM lifetime. Of particular importance concerning the amplification of the mass ion current signal, is the amplification range of the EP 422 amplifier which amplifies the SEM output to an appropriate voltage to be recorded by the QC 422 unit. This was discussed in more detail in Chap. 3.

4.4 Calibration of the DEMS Instrument

The calibration of the DEMS instrument can be simply achieved using a well-defined reaction at the WE, and then measuring the mass ion current of the

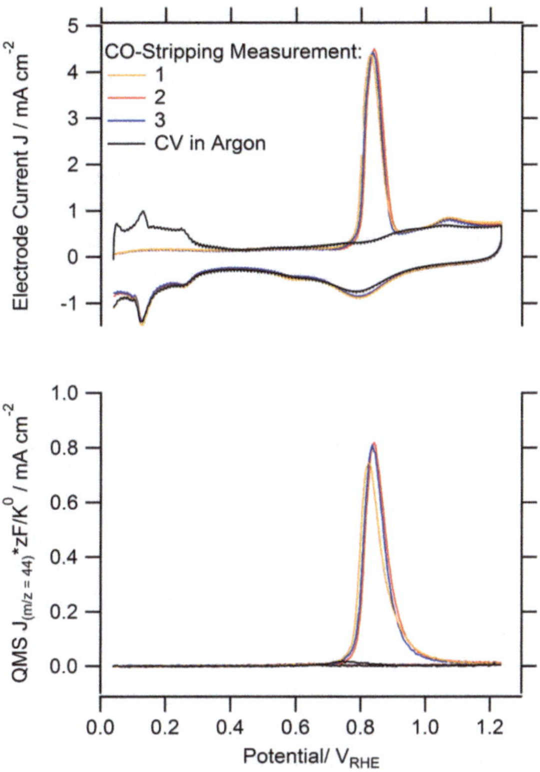

Fig. 4.16 An example of three repeated DEMS CO-stripping measurements on HSAC Supported Tanaka 5 nm Pt catalyst at 10 mV s^{-1} in 0.5 mol dm^{-3} H_2SO_4. The figure demonstrates the precision and reproducibility of the DEMS instrument under stable electrolyte flow cell conditions

corresponding reaction product. The faradaic current, I_F, and I_{QMS} and mass ion current are then related by a constant, K^*, whereby:

$$K^* = \frac{zI_{QMS}}{I_F}.$$

For carbon dioxide, K^*may be most straightforwardly obtained using the CO-Stripping measurement procedure. An example of three calibrating CO-Stripping measurements, including background CV in Argon saturated electrolyte on a HSAC-supported Pt catalyst are shown in Fig. 4.16.

The CO-Stripping procedure is later discussed in more detail elsewhere in this thesis (see Sect. 5.3.1) but briefly, the measurement involves adsorbing a saturated layer of CO_{ads} species on Pt and then oxidising this layer to carbon dioxide during a CV. This produces an oxidative electrode current peak corresponding to the CO_{ads} oxidation charge which is determined by the electrochemical active surface area (ECSA) of the Pt surface. The faradaic electrode charge arising from the oxidation of the CO_{ads} layer on the WE, is therefore related to the mass ion charge (m/z = 44) by a certain constant, such that:

$$K^* = \frac{zQ_{m/z=44}}{Q_F}.$$

where z = the number of electrons, Q_F = faradaic current charge and $Q_{m/z\,=\,44}$ = integrated mass ion current response. In order to appropriately integrate these peaks the background CV for the Pt catalyst in Ar saturated solution must be subtracted from the CO-Stripping CV, and the $I_{m/z\,=\,44}$ background signal (where no electrode reaction occurs) must be subtracted from the measured response to the CO_2, as was the case for CV and MSCV integration in Fig. 4.17. The evaluation of the K^* from the CO-Stripping measurement, however, is an over simplification because K^* contains two terms: (1) the DEMS cell collection efficiency, N, and (2) the calibration constant for the relevant species of the QMS, K^0, so that in general:

$$K^* = \frac{K^0 N}{F}$$

While the parameter K^0 is independent of the electrochemical cell; K^*, must be measured each time in order to determine N. The collection efficiency therefore is dependent upon the DEMS electrochemical cell parameters, as we saw previously in Sect. 4.1.2.4. If K^0 is known, then K^* may be measured, from which a value for N is obtained for each set of experiments. The exact value of N is dependent upon the flow rate of electrolyte over the membrane, which in turn is determined by the cell geometry, thin-layer thickness and the permeability of the analyte species through thc membrane.

Typical values for N for the dual-cyclone thin-layer cell are 0.3 at a flow rate of 5 μ l s^{-1}. Although K^* may be used, it is limited in use because each mass ion current channel (i.e. m/z ratio) must be calibrated for each experiment. In contrast K^0 can be determined for a range of molecules independent of the DEMS cell parameters. Therefore, if K^0 is known, then N can be measured using a single CO-Stripping experiment, thus calibrating the DEMS cell for CO_2. If the collection efficiency for several species are very similar (such as dissolved gases within experimental uncertainty), then this calibration constant can be extended to those species. However, the collection efficiency for volatile organic compounds (VOC) can be expected to be much lower than that for a gaseous species such as carbon dioxide. In such cases, however, the collection efficiencies of VOC species could be pre-determined and related to the CO_2 collection efficiency by coefficients (for various flow rates), which would allow the calibration of multiple species to be achieved for each experiment using a single CO-Stripping measurement. This will, however, need to be confirmed experimentally.

4.5 Calibration of the QMS

Correct calibration of the QMS is very useful in order to perform quantitative DEMS studies that correlate measured mass ion currents with the faradaic electrode current. This may be achieved by leaking a known flux of pure gas into the DEMS vacuum system, and measuring the corresponding mass ion current

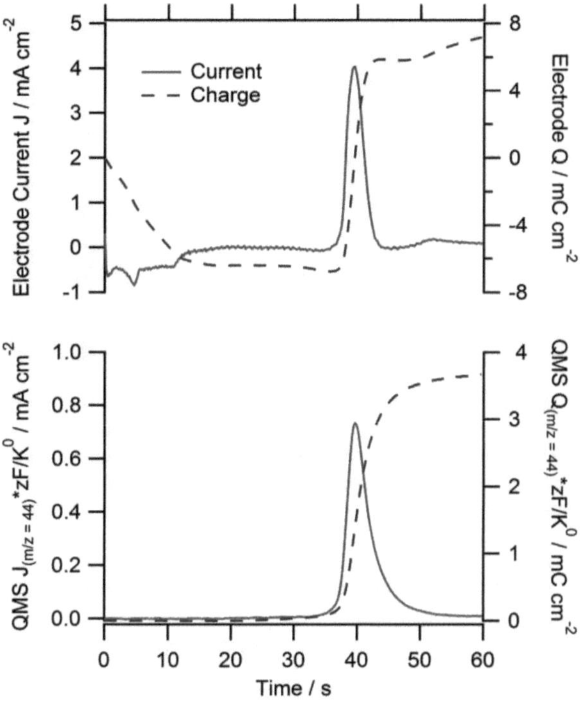

Fig. 4.17 An example background subtracted and integrated CO-Stripping measurement for a 50 μg_{Pt} cm^{-2} sample of HSAC supported Pt Tanaka 5 nm catalyst. By subtracting the background CV in Argon saturated electrolyte, capacitive and surface reaction contributions to the charge are removed, giving the charge associated with the oxidation of the CO_{ads} layer only

response, as previously described in Ref. [6]. The mass ion current, I_i, recorded by the QMS is directly proportional to the flux of species, J_i, into the vacuum chamber, related by a calibration constant, K^0.

$$I_i = K^0 J_i.$$

The gas flux is most easily measured using a calibration volume, V_c, whereby the flux of species, J_i entering a vacuum from V_c is described by:

$$J_i = \frac{V_c}{RT}\left(\frac{dp_c}{dt}\right)$$

where, R is the ideal gas constant (0.08 mbar l s^{-1}), T is the calibration volume temperature ($\sim$298 K) and p_c is the pressure (mbar) of the calibration volume (V_c). It therefore follows that:

$$I_i = K^0\left(\frac{V_c}{RT}\right)\left(\frac{dp_c}{dt}\right).$$

By plotting I_i against $\left(\frac{dp_c}{dt}\right)$ a slope of is obtained.

Because V_c, R and T are known or can be easily measured, the QMS calibration constant, K^0, can be acquired. The exact value of K^0, however, is dependent on a

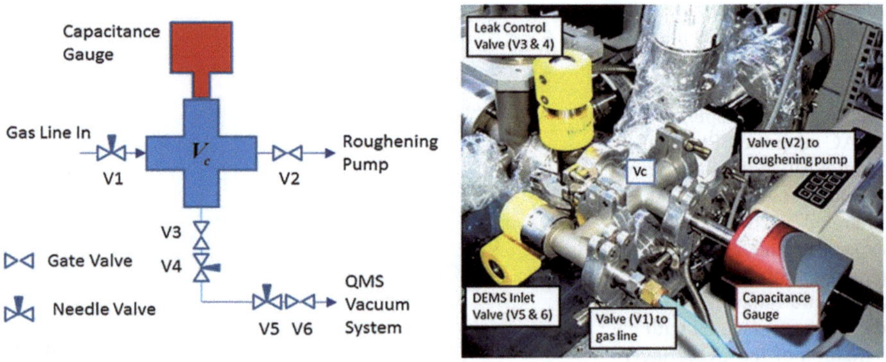

Fig. 4.18 A schematic and photograph of QMS calibration volume setup. The QMS calibration setup consists of a small vacuum chamber, capacitance pressure gauge and numerous valve connections connecting the calibration gas line, roughening pump and QMS vacuum system. The calibration gas is leaked from the vacuum chamber in the QMS whilst both the calibration volume pressure decay and relevant mass ion currents are monitored

combination of the vacuum system, all QMS settings, and of course the particular molecule of interest.

An illustration of the QMS calibration setup is depicted in Fig. 4.18, and consists of a capacitance gauge, calibration gas line and roughening pump which are each attached to a calibration volume that is connected to the DEMS vacuum system via a control valve. The calibration volume of 40 cm^3 was precisely measured using the method described in Ref. [11].

A typical QMS calibration procedure commences as follows:

1. The calibration volume is first evacuated using the roughening pump.
2. Gate valve 2 is then closed, and the calibration gas is slowly leaked in via needle valve V1 into the evacuated volume until a pressure of $\sim$90 mbar is obtained.
3. The needle valve V4 is then partially opened (ca. 5 1/5 turns) the degree to which determines the leak rate of the calibration volume in preparation for the instrument calibration. At this point gate valve V3 remains closed.
4. The gate valve V6 and pin valve V5 of the DEMS vacuum system inlet valve should then be opened slowly until fully opened.
5. Once the calibration measurement and software setup (described in Sect. 3.6.2) is acquiring, displaying and recording the pressure and mass ion current data, the calibration measurement is ready to be performed.
6. In order to allow the flux of gas to enter the vacuum system, gate valve V3 is opened rapidly, and the calibration gas contained within V_c flows into the DEMS vacuum system.

The response of the corresponding mass ion currents is then measured in conjunction with the pressure decay inside the calibration volume using the QMS calibration measurement setup and custom Labview calibration software. An

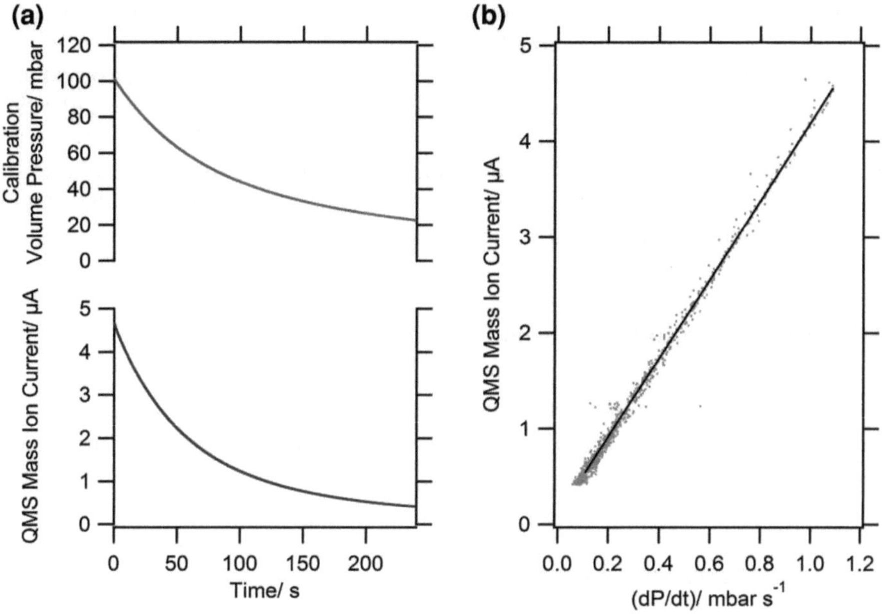

Fig. 4.19 An example QMS calibration measurement showing the calibration volume pressure decay and mass ion current response (**a**) and the corresponding relationship between gas flux and mass ion current (**b**). The gas flux is determined by the conductance of the valves and the pressure difference between the calibration volume and DEMS vacuum system

Table 4.3 List of QMS calibration constants, K^0, for a number of gaseous species obtained for this DEMS instrument.

	Molecule	m/z	SEM voltage/ V	
			1,300	1,600
QMS calibration constant, K^0/C mol^{-1}	Ar	40	0.18	2.15
	O_2	32	0.12	1.61
	H_2	2	0.06	0.78
	CO_2	44	0.19	2.65
	CO	28	0.16	2.03

QMS parameters: Emission current 1 mA, Ion ref. 85 eV, Cathode 70 eV, Focus plate 14.0 V, Extraction 250 V, Field Axis 10.50 V, Deflection 200 V and Resolution 30. Measurements are reproducible to within ~0.005 C mol^{-1}

example measurement is shown in Fig. 4.19a along with the respective calibration curve b.

The mass ion current of the relevant mass ion current channel (m/z ratio) is linearly proportional to the flux into the DEMS vacuum system, and a very linear calibration curve can be obtained giving a value K^0. A table of typical QMS calibration values for a number of gaseous species are given in Table 4.3. The calibration constants are of course dependent on the gaseous species along with the exact settings of the QMS and nature of the vacuum system, and are of course only relevant to the this particular DEMS instrument.

This calibration technique may also be extended to volatile organic compounds such a methylformate, however, an injection port and heating of the calibration volume would be necessary in order to introduce the sample and prevent the condensation of the species onto the inner walls of the calibration volume, as highlighted in Ref. [6]. There are of course other methods of QMS calibration by using, for example, gas quantities described previously in Ref. [12].

4.6 Further Considerations

The following subsections contain descriptions of two further areas that must be considered in DEMS experimentation: measurement error and instrument maintenance.

4.6.1 Measurement Error

All measurements are subject to errors that contribute to the uncertainty of a measured value. It is therefore important to be able quantify errors in the DEMS measurement, or any assumptions in data treatment in order to distinguish between true experimental observations and those that arise from uncertainties to avoid the possible misinterpretation of data.

In experimentation where the DEMS instrument is used for purely qualitative observations, then any quantitative errors are clearly less of an issue and reproducibility of the qualitative observation is. However, when the DEMS instrument is used in a quantitative manner (which is far more interesting), then the sources of error for the DEMS instrument are dependent on a number of contributing factors which we need to be aware of. In order to quantify the measured QMS mass ion current, and calculate the equivalent faradaic current for an electrode, the following equation is used:

$$I_{F_{QMS}} = \frac{I_{QMS} zF}{K^0 N}.$$

The error in the equivalent faradaic current calculated from the QMS mass ion current contains an error that are propagated from other experimental uncertainties, namely K^0, N and I_{QMS} as given in the following equation:

$$\delta I_{F_{QMS}} = \left| I_{F_{QMS}} \right| \cdot \sqrt{\left(\frac{\delta I_{QMS}}{I_{QMS}}\right)^2 + \left(\frac{\delta K^0}{K^0}\right)^2 + \left(\frac{\delta N}{N}\right)^2}.$$

These experimental uncertainties, namely δI_{QMS}, δK^0 and δN are defined in the following manner. The uncertainty in the calibration constant, K^0, can be easily accessed by performing multiple calibration measurements and taking the standard deviation of the measurements. The error in DEMS cell collection efficiency, *N*, may be obtained from performing multiple CO-Stripping measurements prior to an experiment, and represents the most significant error, typically $\sim 10\ \%$ or less. The error in the measured I_{QMS}, however, is slightly more difficult to define because the absolute value relies on a background correction of the mass ion current signal which can be subject to gross errors if the background is ill-defined (or unstable) and subtracted improperly. In the calculation of the DEMS instrument error bars in this thesis, the mass ion current uncertainty, δI_{QMS}, is defined as the QMS detection limit which is 2.5 multiplied by the standard deviation of the mass ion current signal noise.

All these errors are of course susceptible to gross errors. If the I_{QMS} background current is unstable or ill-defined and subtracted incorrectly, then the background corrected values are then subject to a potentially large systematic error, and obviously in such cases the interpretation of experimental data should obviously be treated with caution. Similarly, the sensitivity of the DEMS cell collection efficiency, *N*, on the electrolyte flow rate can also lead to large errors if the electrolyte flow fluctuates significantly or is momentarily disturbed or impeded effecting measurement results. These gross errors, however, should be easily spotted by an experienced DEMS user. Finally, the sensitivity of the QMS will to some extent deteriorate and drift through use (changing K^0 for the various species) as the ion source filaments and SEM wear out. The calibration constant therefore should be assessed periodically, or when the instrument has not been used for some time.

4.6.2 Maintenance

In order to maintain the DEMS instrument and (hopefully) provide many years of reliable use, the following aspects of maintenance concerning the electrochemical flow cell, membrane interface and vacuum system should be considered.

Regarding the DEMS electrochemical flow cell, all PTFE and Kel-f parts should be regularly cleaned and boiled in line with the standard PTFE electrochemical RDE setup protocol. In certain circumstances it is not strictly necessary to reconstruct the DEMS cell for each daily measurement and the flow block can be left attached to the membrane interface and vacuum flange; however, the thin-layers of the cell should be extensively purge with Millipore water after, and prior to experimentation. It is also important to monitor for corrosion on the DEMS cell vacuum flange and steel frit. Another aspect of the electrochemical flow setup concerning maintenance, is the syringe pump and flow cell lines. The performance of the syringe will degrade through use, particularly as it is generally in contact with acid electrolyte which hardens the rubber seal over time and will likely be the cause of inconsistent flow, although deterioration in the sealing of the flow line connections can also be problematic. Furthermore, the thread of the syringe pump mechanism is also prone to being damaged if the mechanism of the pump is subjected to any strong resistance.

The microporous PTFE membrane material is in general very durable, and can in principle by used for many days of experimentation without adversely affecting performance provided that the DEMS cell is not dismantled. The PTFE cell gaskets meanwhile are re-usable, but the stainless steel metal frit and flange are prone to slight corrosion.

In general, the vacuum system requires the same maintenance procedures typical to vacuum hardware, such as, the monitoring of the conditions of the turbo pumps (by sound and the current drawn), particularly that which pumps the 1st pumping stage (which operates at rather high vacuum pressure), and subsequently replacing their oil reservoirs on a yearly basis. Of particular concern regarding the DEMS instrument system is, that an appreciable amount of aqueous acidic electrolyte evaporates into the vacuum system. Whilst the turbo-pumps appear to be largely unaffected by this (although only time will tell), the roughening pumps are prone to being damaged quickly. This is because the aqueous acid electrolyte condenses in the roughening pump oil, which not only changes the composition of the oil but also begins to corrode the pumps mechanical parts and seals. Consequently, it is recommended that after using the DEMS instrument the roughening pumps are operated with the gas ballast valve open for ~1–2 h to assist in purging the aqueous contaminants from the vacuum pump oil. Although helping, the gas ballast will not completely resolve the problem. The roughening pump oil should therefore be periodically replaced, every 2–3 months depending on the use of the DEMS, particularly the Trivac D16B pump which backs the 1st pumping stage, and thus pumps the most acid vapours.

4.7 Summary

The purpose of this chapter was to present an understanding of the practical aspects of the DEMS instrument, concerning the influence and optimisation of experimental parameters, the settings of the QMS and the performance of the membrane whilst to also document the instrument calibration procedures.

As a result of the narrow capillaries and thin-layer of the electrochemical cell design, potential control of the WE cannot be achieved without modifying the electrochemical setup. The origin of the potentiostat oscillation in this cell was too small of a solution resistance (<1 Ω) between WE and the effective Luggin-capillary. The potentiostat was consequently stabilised by adding a resistance of ~30 Ω to the WE connection, which effectively acts as an artificial solution resistance. This rather high artificial solution resistance was then compensated for using a NID to provide a more acceptable ~10 Ω uncompensated resistance in 0.5 mol dm^{-3} H_2SO_4 to minimise systematic errors, without impinging on the potentiostat circuitry.

The performance of the DEMS instrument was found to be strongly dependent on the electrochemical cell variables, particularly the electrolyte flow rate. The diffusion limited current for the ORR, HOR and CO-bulk oxidation was observed to scale with the volumetric flow rate by a power of 1/3. Of more importance to DEMS, however, is the influence of the flow rate on the QMS response time, the degree of correlation between the faradaic electrode current and the mass ion current, as well as the DEMS cell collection efficiency. By systematically assessing the effect of flow rate on these parameters, optimised flow rates were found to be between $\sim$4 and 10 μl s^{-1} at potential scan rates $\leq$20 mV s^{-1} giving DEMS cell collection efficiencies in the range of 0.4 to 0.1.

The performance of prospective microporous membrane materials was characterised in terms of its permeability and selectivity. The permeability of the membrane was assessed using a test setup which measured the total gas fluxes through the wetted membrane when exposed to a vacuum. Typical total gas fluxes for suitable DEMS membranes were in the range of 0.1 and 0.2 mbar l s^{-1} at room temperature. In terms of selectivity, the instrument is very capable of detecting dissolved gaseous electrochemical products but is not suitable for the in situ detection of polar or ionic electrochemical reaction species. If there is a desire in the future to use alternative membrane materials, or non-aqueous electrolytes then these combination should be performed on a test setup, rather than risk the flooding of the DEMS vacuum system.

The optimisation of the QMS parameters is crucial toward maximising the signal-to-noise ratio of the DEMS instrument and henceforth the detection limit of the instrument. An overview of the effect of QMS parameters on the measured mass ion current was presented in order to aid in the understanding and tuning of the QMS. The majority of the noise was found to arise from the amplification of the SEM signal by EP 422, and it is of particular importance that this is set manually in the appropriate amplification range.

The calibration of the DEMS instrument meanwhile involves the determination of the DEMS cell collection efficiency, and calibration of the QMS. In principle, the calibration of the DEMS collection efficiency should be performed prior to experiment using CO-Stripping. The QMS calibration values for a certain species should be measured each time a QMS parameter is changed, and occasionally thereafter to monitor changes in the QMS performance. The sensitivity of the instrument, however, is defined by the DEMS cell collection efficiency, and the signal-to-noise ratio of the QMS.

The sources of DEMS error include those that are common to all electrochemical half-cells, such as the systematic error introduced by uncompensated solution resistance, impurities, incomplete gas purging of the electrolyte and shifts in the potential of the RE. Of more specific concern for DEMS are errors that can arise from unstable electrolyte flow rate which can have effect on the diffusion limited current, the uncertainty in DEMS calibration constant, QMS response time and correlation. Finally, the DEMS instrument also requires the standard maintenance routines for vacuum equipment, however, the condition of the oil in the roughening pumps should be monitored particularly carefully due to the detrimental effect that the traces of acidic electrolyte which pass through the membrane interface have on the oil composition and corrosion of the pumps mechanical parts.

References

1. Technical Note 200 Getting to know your potentiostat: Potentiostat stability considerations. 2010, Princeton applied research
2. Britz D (1978) iR elimination in electrochemical cells. J Electroanal Chem 88(3):309–352
3. Electrochemistry—Application note n°4: Potentiostat stability mystery explained. 2010: BioLogic Science Instruments
4. Lamy C, Herrmann cc (1975) A new method for ohmic-drop compensation in potentiostatic circuits—stability and bandpass analysis, including effect of faradaic impedance. J Electroanal Chem 59(2):113–135
5. Brett CMA, Brett AMO (1993) Electrochemistry: principles, methods, and applications. Oxford University press, Oxford
6. Baltruschat H (2004) Differential electrochemical mass spectrometry. J Am Soc Mass Spectrom 15(12):1693–1706
7. Wolter O, Heitbaum J (1984) Differential electrochemical mass-spectroscopy (DEMS)—a new method for the study of electrode processes. Berichte Der Bunsen-Gesellschaft-physical chemistry. Chem Phys 88(1):2–6
8. Operating instructions, quadrupole mass spectrometer system, QMG 422. Vol BG 805 983 BE: Pfeiffer vacuum GmbH
9. Quadstar 32-bit, Application Program for Quadrupole Mass Spectrometers QMS 422 and QMS 200.: Pfeiffer Vacuum GmbH
10. Operating instructions, quadrupole analyzer QMA 125: Pfeiffer vacuum GmbH.
11. Kendall BRF (1974) Apparatus for measuring the internal volumes of vacuum systems. J Vac Sci Technol 11(3):610–611
12. Dobrozemsky R (1990) Calibration of vacuum systems by gas quantities. Vacuum 41(7–9):2109–2111

Chapter 5
Methanol Oxidation on HSAC Supported Pt and PtRu Catalysts

In this first demonstration of this DEMS instrument research application, the study of the methanol oxidation reaction (MOR) on HSAC supported -Pt and -PtRu catalysts is revisited. The objective of this study was to not only demonstrate the research capabilities of the DEMS instrument constructed in this thesis but to also examine the electroanalytical techniques that are commonly employed in order to study and assess the MOR reaction and electrocatalyst activities using RDE. The results presented here, however, do not simply repeat previous observations but elucidate the contrasting potential dependent conversion of the MOR to CO_2 on Pt and PtRu catalysts. Three-dimensional voltammetry is furthermore applied for the first time combined with DEMS and utilised to describe the MOR, encapsulating the potential, current and time relationship of the reaction system within a single contour plot. The observations demonstrate that the MOR on the Pt catalyst correlates directly with the conversion of methanol to CO_2; however, this relationship was not apparent in the PtRu catalyst signifying a key difference in the MOR mechanism and rate determining step.

5.1 Introduction

Direct methanol fuel cell (DMFC) CEEC devices possess the obvious advantage over compressed hydrogen fuelled PEMFCs that the liquid methanol fuel can be easily transported and stored using existing infrastructure. Unfortunately, even with rather high electrocatalyst loadings (4 mg_{Pt} cm^{-2} [17]) and elevated temperature (>100 °C) DMFCs only achieve relatively poor power densities (100–250 mW cm^{-2} [17]) in contrast to hydrogen fed PEMFCs (700 mW cm^{-2} with catalyst loadings of 0.6–0.8 mg_{Pt} cm^{-2} [1]). This comparatively poor power density is a result of the critical technical obstacle concerning DMFC development, namely the large operating electrode overpotentials required for both the

S. J. Ashton, *Design, Construction and Research Application of a Differential Electrochemical Mass Spectrometer (DEMS)*, Springer Theses,
DOI: 10.1007/978-3-642-30550-4_5,

MOR at the anode, and oxygen reduction reaction (ORR) at the cathode [18]. Despite that these limitations inhibit DMFC application in transportation, passive DMFCs with low power densities (6–27 mW cm^{-2} with loadings of 4–16 mg_{Pt} cm^{-2}) operating at ambient temperature (<40 °C) are nonetheless rather appealing to the portable electronic device market, where they possesses a number of potential advantages over Li ion battery technology. For instance, the theoretical specific energy density of the liquid methanol fuel is 6,000 Wh kg^{-1} which is some ten times higher compared to the Li ion battery at 600 Wh kg^{-1} [17]. Small portable DMFCs therefore, not only offer consumers the possibility for their portable electronic devices to be used for longer, but also allows instantaneous recharge (or rather refill), as opposed to the hours required to recharge a battery. Regardless of application, in order for the DMFC development targets to be met a reduction in the precious metal electrocatalyst loading is essential [17]. Although the large overpotential for the ORR (common to all PEMFCs) and methanol crossover (diffusion of methanol from anode to cathode) contribute significantly to the performances losses of the DMFC, the overpotential required for the MOR at the anode is of even greater concern. Extensive research efforts have consequently focused on the study of the MOR, particularly in electrochemical half-cells in an endeavour to understand the fundamental aspects of the reaction mechanism and the relationship between electrocatalyst structure and activity [4–6, 19, 20] in order to aid in the development of more active and cheaper electrocatalyst materials. Even so, rather expensive PtRu based electrocatalysts are still by far the most active anode materials for the MOR, although there is certainly an incentive to develop alternative non-Pt based electrocatalyst materials [21].

5.1.1 Background

The complete electrochemical oxidation of methanol to carbon dioxide is described by the following overall reaction formula: .

$$CH_3OH + H_2O \xrightarrow{E} CO_2 + 6H^+ + 6e^- \quad E^0 = 0.046\ V$$

Although the low standard redox potential for the MOR is theoretically favourable for a CEEC device, reasonable MOR rates are unfortunately not achieved using even the most active Pt-based electrodes until a positive overpotential of several hundred millivolts is applied, reducing the practical performance of the DMFC device. This overpotential is necessary because the complete MOR proceeds via multiple reaction steps involving partially oxidised intermediates, the most significant of which is the strongly adsorbed CO_{ads} species that poisons the Pt electrocatalyst surface inhibiting further methanol adsorption, as described by:

$$CH_3OH \xrightarrow[Pt]{E_1} Pt - CO_{ads} + 4H^+ + 4e^-$$

The oxidation of the CO_{ads} species on Pt in turn requires the dissociative adsorption of water to form OH_{ads} species, as given by the following formula:

$$H_2O \xrightarrow[Pt]{E_2} Pt - OH_{ads} + H^+ + e^-$$

The oxidation of the CO_{ads} species may then proceed via the bifunctional reaction of adsorbed CO_{ads} and OH_{ads} species:

$$Pt - CO_{ads} + Pt - OH_{ads} \xrightarrow{E_3} 2Pt + CO_2 + H^+ + e^-$$

The adsorption of OH_{ad} is therefore critical to the oxidation of CO_{ads} and in turn the MOR; however, OH_{ads} species are not formed on pure Pt surfaces until much higher electrode potentials i.e. $E_2 \gg E_1$. As a consequence of the self-poisoning nature of the MOR and the limitation imposed by the slow adsorption of OH_{ads} on Pt necessary for the oxidation of poisoning CO_{ads} species, pure Pt materials are generally considered rather poor MOR catalysts. Certain alloys of Pt such as PtRu have, however, proved to exhibit superior MOR activity in comparison to pure Pt. On PtRu this improvement is attributed to the adsorption of OH_{ads} species on Ru surface atoms at lower electrode potentials, and is similarly described by the formula:

$$H_2O \xrightarrow[Ru]{E_4} Ru - OH_{ads} + H^+ + e^-$$

The consequence of the adsorption of OH_{ads} on Ru at lower electrode potentials, i.e. $E_4 < E_2$, is that the oxidation of CO_{ads} on Pt can then proceed by the bifunctional reaction given by:

$$Ru - OH_{ads} + Pt\text{ - }CO_{ads} \xrightarrow{E_5} Pt + Ru + CO_2 + H^+ + 2e^-$$

The poisoning CO_{ads} species are therefore oxidised at lower overpotentials, liberating the Pt surface and allowing methanol to adsorb and reaction proceed at lower potentials, where $E_5 \sim E_4$. The combined action of the Pt and Ru atoms is often referred to as being co-catalytic, whilst the CO_{ads} oxidation mechanism can be considered bifunctional in both senses of the term i.e. the reaction proceeds via two different adsorbed functional groups (CO_{ads} and OH_{ads}) involving two different adsorption sites (Pt and Ru).

The poisoning CO_{ads} species, however, is not the only partial MOR intermediate involved in the complete MOR to carbon dioxide. A number of detailed mechanisms have been proposed in the scientific literature, some of which are quite elaborate [22, 23] while others more straightforward [15]. The MOR mechanism presented in Fig. 5.1 is of the latter straightforward variety and is based on MOR products and predominantly adsorbed surface species, observed using in situ methods such as DEMS and FTIR. For example, DEMS has been previously used to measure the consumption of methanol [15], and/or the production of methylformate ($HCOOCH_3$) and carbon dioxide [8, 9, 11, 13, 15,

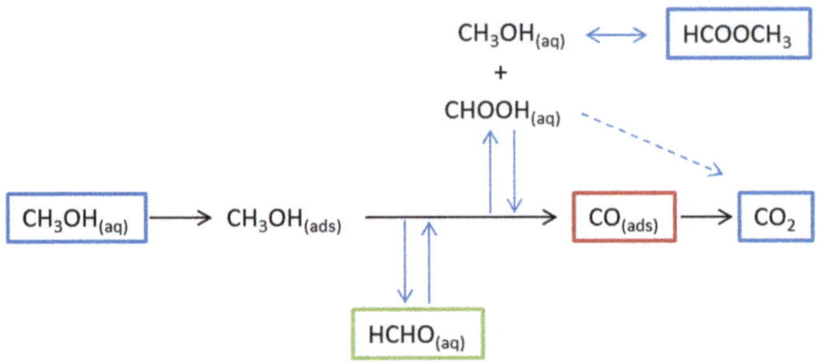

Fig. 5.1 A diagram of the electrochemical MOR pathway. Methanol (CH_3OH), methylformate ($HCOOCH_3$) and CO_2 (*blue*) have been observed using DEMS, CO_{ads} (*red*) using FTIR/SPAIRS whilst formaldehyde (HCHO—*green*) has been found (ex-situ) using fluorescence spectroscopy. In order for methanol to be completely oxidised to CO_2 all desorbed partially oxidised MOR intermediates must be further oxidised

24], whereas surface spectroscopy techniques such as FTIR [25] or single-potential alteration infrared spectroscopy (SPAIRS) [4] have primarily observed CO_{ads} species along with some additional absorption bands that have been assigned to various adsorbed species such as COH_{ads} and $CHOO_{ads}$. Formaldehyde yields of the MOR have meanwhile been observed ex-situ using fluorescence spectroscopy [26, 27].

At ambient temperature (<40 °C) the MOR is initiated via the dissociative adsorption of methanol on the Pt surface and may proceed via two pathways termed direct and in-direct. In the direct pathway, the reaction proceeds directly to CO_{ads} which must then be oxidised to evolve CO_2 yielding the full six electrons. Alternatively, methanol may only be partially oxidised to weakly adsorbed intermediates that can desorb in the form of either formaldehyde [26] or formic acid [8, 9, 11, 13, 15, 24], yielding only two or three electrons respectively. The desorbed formaldehyde and formic acid partial MOR products must therefore re-adsorb onto the electrocatalyst surface in order to oxidise fully to CO_2, a process that can be termed the in-direct pathway. To further complicate the MOR system, these desorbed partial MOR products may also undergo chemical reaction with species in the electrolyte. For example, it has been noted that formaldehyde largely exists in the form of methylene glycol ($H_2C(OH)_2$) in acid electrolyte whose oxidation may also proceed via two pathways [28, 29]. Similarly, formic acid may again, oxidise via two pathways or alternatively undergo an acid catalysed esterification reaction with methanol in solution to form methylformate (as observed using DEMS [24]) which electrochemically oxidises to yield CO_2 and methanol [30].

Overall, in order for the complex MOR processes to be considered complete, all desorbed partial MOR products must re-adsorb and oxidise completely to CO_2 in order for full conversion to be achieved, yielding the full six electrons per adsorbing methanol molecule.

5.1.2 *Motivation*

In light of the extensive research efforts that have already been exhausted on the study of the MOR on various Pt and PtRu based catalysts under numerous experimental conditions, what can yet another electrochemical half-cell study on the MOR hope to achieve. Well, there are three main aspects to the motivation for this study. The first actually concerns the performance of DEMS instrument rather than the study of MOR itself. It is more than reasonable to choose a topic which has already been relatively extensively studied using DEMS [8, 10, 12, 13, 15, 31] in order to first demonstrate and assess the research capabilities of the instrument constructed as part of this thesis. This, however, does not mean to say that this investigation simply repeats the observations of previous DEMS studies. The second aspect of the motivation for this study was to evaluate, using DEMS, the electroanalytical techniques often employed using standard RDE methods to characterise and compare the electrocatalytic activity of prospective MOR catalysts. In doing so, this study will highlight that there is no consistently employed method to evaluate and compare the activity of such materials toward the MOR, with many studies in particular focusing on rather high electrode potentials >>0.5 V_{RHE}. The final aspect, concerns whether new insights can be offered into the MOR by, for example, applying more sophisticated electroanalytical methods such as three-dimensional voltammetry combined with DEMS, despite all that has been done previously.

5.2 Experimental

Each electrocatalyst sample (Tanaka Pt 5 nm 50.6 % wt. Pt on, and Tanaka 78 % wt. PtRu (1:1) catalysts supported on EC300 HSAC) was prepared by ultrasonically dispersing the catalyst powder in 50 ml of Millipore Milli-Q water. The measurement sample was prepared by pipetting 20 μL of the aqueous catalyst suspension onto a mirror polished 5 mm Ø glassy carbon WE substrate to give an electrode loading of 50 μg_{metal} cm^{-2}. All measurements were performed in 0.5 mol dm^{-3} H_2SO_4 electrolyte prepared in Millipore Milli-Q water (<18.3 M Ohm cm^{-1}, TOC < 5 ppb) using concentrated H_2SO_4 (Normaton; Merck, Germany). The MOR reaction measurements were meanwhile performed in the H_2SO_4 acid electrolyte containing a methanol concentration of 1.0 mol dm^{-3} (Proanalysis, Merck, Germany).

The DEMS instrument setup consists of a dual-cyclone thin-layer electrochemical flow cell, 0.03 μm microporous PTFE membrane (Gore) interface, a 3-stage differentially pumped high vacuum system and QMS (Pfeiffer Vacuum QMA 422) as was described previously in Chap. 3. The electrochemical and membrane interface thin-layer compartments are formed and sealed tight using 250 μm thick PTFE gaskets resulting in volumes of ~4 and 7 μL in the WE and membrane interface compartment, respectively. The CE was positioned in a side-compartment located to the side of the flow cell inlet, whilst the RE is located in a sealed compartment positioned on the outlet of the flow cell. An electrolyte flow rate of 5 μl s^{-1} through the thin-layer flow cell was maintained using a KDS 200 precision syringe pump, located on the flow outlet, withdrawing electrolyte from an Ar (purity 5.0) gas purged reservoir. In order to avoid potentiostat oscillations that arise because of high CE and RE solution resistances to the WE, and low solution resistance to the Luggin-capillary in dual thin-layer flow cell, a stabilising resistor (~30 Ω) was added in series to the WE connection and compensated for using a custom built NID. The resulting uncompensated solution resistance, typically ~10 Ω, was then corrected for during post measurement data treatment.

The faradaic current arising from the complete MOR reaction to CO_2, I_{CO2}, was calculated from the mass ion current, $I_{m/z\,=\,44}$, using the following equation:

$$I_{CO_2} = \frac{I_{m/z=44} \cdot 6F}{K^0_{CO_2} N_{CO_2}}$$

The QMS calibration constant, K^0, for CO_2 at m/z = 44 was 2.95 C mol^{-1}, the DEMS cell collection efficiency, N, for CO_2 was determined experimental using CO-stripping prior to the experiment, typically 0.3 at an electrolyte flow rate of 5 μl s^{-1}. The WE surface is continuously supplied with fresh 1.0 mol dm^{-3} methanol in H_2SO_4 acid electrolyte, and a total of 6 electrons are assumed to arise for each CO_2 observed. The background current signal for $I_{m/z\,=\,44}$ was subtracted post measurement, whereby the background was defined as the stable $I_{m/z\,=\,44}$ signal observed when no electrode reaction took place. The I_{CO2} was furthermore used to determine the conversion (expressed as a fraction, also commonly referred to as current efficiency in previous DEMS papers) of the MOR to CO_2 using the following equation:

$$Conversion = \frac{I_{CO_2}}{I_F}$$

In addition to $I_{m/z\,=\,44}$, the $I_{m/z\,=\,60}$ corresponding to methylformate (CHO-OCH_3) was also measured during the MOR measurements presented in this work. The $I_{m/z\,=\,60}$ signal, however, was (unfortunately) not calibrated in this work and all observation of $I_{m/z\,=\,60}$ are purely qualitative. Consequently, the results and discussion of the $I_{m/z\,=\,60}$ measurements has been largely omitted in this study; however, the data is presented in Appendix A for future reference.

All electrode potentials meanwhile are given versus the RHE determined using the HOR/HER in hydrogen saturated electrolyte determined using a polycrystalline Pt electrode prior to experimentation.

5.2.1 Measurement Procedure

The catalyst samples were initially characterised using CV at various scan rates in Argon saturated solution, by cycling the potential between 0.05 and 1.15 V_{RHE} for the Pt catalyst, and 0.05–0.85 V_{RHE} for the PtRu catalyst. A CO-Stripping measurement was then performed in order to determine the ECSA of the catalyst samples prior to the study of the MOR. Following the CO-Stripping measurement, the electrolyte was then exchanged with electrolyte containing 1.0 mol dm^{-3} concentration of methanol by switching the electrolyte reservoir and purging the electrochemical cell for ca. 5 min, during which the potential of the WE was maintained at 0.05 V_{RHE}.

The electrochemistry of the HSAC supported −Pt and −PtRu catalysts toward the MOR was first characterised under potential dynamic conditions using CV performed at 10 mV s^{-1} between 0.05 and 1.15 V_{RHE} for the Pt catalyst, and 0.05–0.85 V_{RHE} for the PtRu catalyst. The MOR on the respective catalyst samples was then studied under potentiostatic conditions using a sequence of chronoamperometric measurements. This sequence involved stepping the WE potential from 0.05 V_{RHE} to the desired potential of study, progressively increasing the potential in 50 mV steps from 0.435 to 0.835 V_{RHE} in the case of the Pt catalyst, and 0.435–0.735 V_{RHE} for the PtRu catalyst, each time holding the potential for 420 s. Between each chronoamperometric measurement three CV were recorded to 'reset' the catalyst and monitor any apparent changes in the CV. By performing this series of multiple chronoamperometric measurements between the potential region of interest, three-dimensional plots can be constructed to encapsulate the time, potential and current relationship using Wavemetrics© Igor Pro 6.21 *contour plot* graph function. This is most easily achieved by arranging the data into single X, Y and Z waves and then building the contours using ca. ×16 interpolation. The 3-dimensional plot matrix was then constructed from the contours using the *Contour Z* function.

5.3 Results and Discussion

5.3.1 Electrochemical Surface Area Determination

In this study on the MOR, the ECSA of the HSAC Pt and PtRu catalysts were first determined using CO-Stripping, whereby a saturated layer of CO_{ads} species is adsorbed at a potential of 0.05 V_{RHE} which is subsequently oxidised during a LSV. In DEMS this serves two purposes: firstly, it can be used to calibrate the DEMS cell collection efficiency (N_{CO2}), and secondly, it is a measure of the ECSA of the catalyst sample. The results from a typical DEMS CO-Stripping measurement of the HSAC Tanaka Pt 5 nm catalyst is given in Fig. 5.2.

Focusing firstly on the CV in Argon saturated electrolyte in Fig. 5.2a, we observe the standard features expected from a pure Pt catalyst. In the positive potential scan the hydrogen under potential desorption (H_{upd}) between 0.05 and 0.35 V_{RHE} is clearly present, followed by the double layer region between 0.35 and ~0.74 V_{RHE} before we begin to oxidise the Pt surface with OH_{ads} species and finally forming PtO as the potential exceeds 0.9 V_{RHE}. In the reverse potential

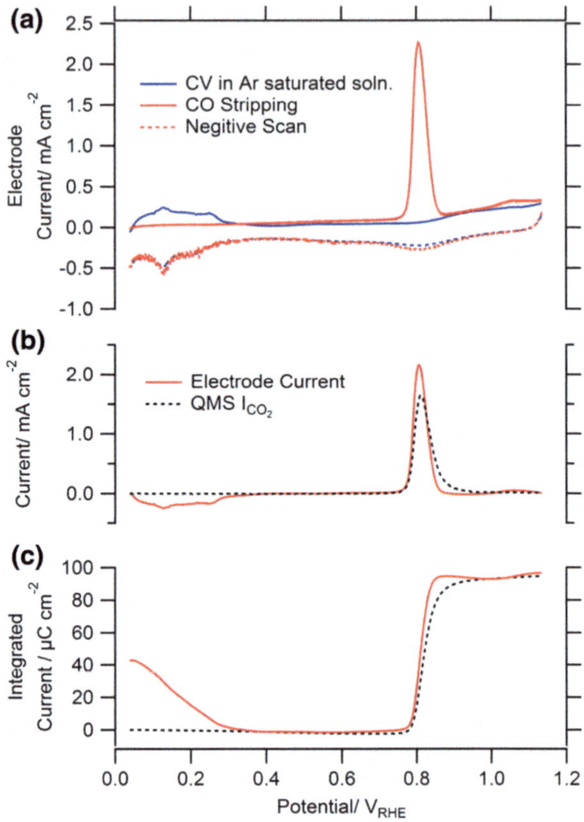

Fig. 5.2 A DEMS CO-stripping measurement of 50 µg cm^{-2} of the Pt 5 nm HSAC supported catalyst. **a** CO-Stripping and background CV in Argon saturated electrolyte, **b** corresponding background subtracted CO-Stripping LSV and I_{CO2}, and **c** charge, performed at 10 mV s^{-1} in 0.5 mol dm^{-3} H_2SO_4. The background subtracted CO-Stripping charge can be almost exclusively attributed to oxidation of the saturated CO_{ads} layer

scan, these Pt oxide species are then reduced, with a visible hysteresis that exhibits a reduction peak at ~0.8 V_{RHE}, which is again followed by a narrowed double layer region before we observe the reduction current of hydrogen adsorption on Pt between 0.35 and 0.05 V_{RHE}. The currents in the CV of the HSAC supported Pt catalyst in Argon saturated electrolyte are dominated by the faradic currents arising from the surface reactions that occur between the Pt and aqueous electrolyte; however, these also contain ill-defined capacitive currents that arise from the whole electrode surface area, including that of the HSAC support and WE substrate.

Prior to the CO-Stripping CV, the WE potential is held at 0.05 V_{RHE} and the electrode surface to exposed to dissolved CO gas that adsorbs onto the Pt surface, displacing the Pt-H_{ads} species and forming a saturated Pt-CO_{ads} layer. This is indicated by the disappearance of the H_{upd} feature in the upward potential scan prior to CO_{ads} oxidation. In the positive potential sweep of the saturated CO_{ads} covered Pt electrode, the current is dominated by double layer capacitive currents until 0.74 V_{RHE} where we observe a sharp increase in current arising from the oxidation of the saturated CO_{ads} layer. Once the CO_{ads} is completely oxidised to

CO_2 freeing the Pt surface, the CV returns to the form observed prior to the CO-Stripping measurement. The CO-Stripping CV, however, still contains capacitive currents arising from the HSAC support, WE substrate and double layer, along with the faradaic Pt surface oxidation currents that are present in the CV in Argon saturated solution. These additional current contributions in the CO-Stripping measurement may be removed by subtracting the CV in Argon, yielding both the CO-stripping and inverse H_{upd} charge, as shown in the LSV of Fig. 5.2b. This is compared with the corresponding faradaic current for I_{CO2} calculated (assuming 2 e^{-}) from the simultaneously measured $I_{m/z\,=\,44}$ using DEMS, where it can be seen that the potential dependent curves of the I_F and I_{CO2} do not perfectly overlap. This broadening in the I_{CO2} is due to the diffusion and mixing of the CO_2 generated at the electrode surface in the DEMS, and is a typical limitation of the technique, nonetheless, the overall calibrated charge in Fig. 5.2c is the same. At this point, it is immediately clear from a simple CO-Stripping measurement why CO_{ads} species that do not begin to oxidise (in a saturated layer) until 0.74 V_{RHE} might inhibit the MOR on a pure Pt surface at DMFC anode relevant potentials <0.5 V_{RHE}.

The significantly earlier onset potential of the oxidation of a saturated CO_{ads} layer on the HSAC supported PtRu catalyst is seen in Fig. 5.3.

The CV of the PtRu catalysts recorded in Argon saturated solution exhibits similar features to those observed on pure Pt catalyst, although the terms H_{upd}, double layer and oxidation/reduction regions are (strictly speaking) only relevant to the CV of pure Pt catalysts, these terms will be used to identify the same potential regions for the PtRu catalyst. A H_{upd}/H_{ads} feature is present in the CV of PtRu between 0.05 and 0.28 V_{RHE}, although in this instance the process is assumed to occur equally on both the Ru and Pt surface atoms based on the CV of pure Ru electrodes [20, 32]. In contrast to the pure Pt catalyst, however, there is no longer a clear distinction between the H_{upd}, double layer and oxidation/reduction regions of the CV. This is a consequence of additional faradaic current arising from the earlier oxidation of the Ru surface atoms forming Ru-OH_{ads} species as early as 0.2 V_{RHE} following H_{upd} and continuing across the double layer potential region where the Pt surface also begins to oxidise. The surface oxide species are subsequently reduced in the downward potential scan, analogous to the pure Pt electrode which is assumed to be reversible. Following the adsorption of dissolved CO gas at 0.05 V_{RHE} on the PtRu catalyst, the H_{upd} feature in the upward potential scan of CO-Stripping sweep is again clearly suppressed suggesting that a saturated layer of CO_{ads} has formed on both the Pt and Ru surface atoms (the adsorption of CO_{ads} on Ru is known from CO-Stripping measurements performed on pure Ru electrodes [20]). The saturated CO_{ads} layer also appears to have suppressed some small oxidative currents between 0.35 and 0.4 V_{RHE} that are present in the CV in Argon saturated solution but not in the subsequent CO-stripping sweep which may suggest that the Ru-CO_{ads} species inhibit the formation of Ru-OH_{ads} in this potential region. The onset of the saturated CO_{ads} layer oxidation on the PtRu catalyst occurs at 0.4 V_{RHE}, ~350 mV earlier than that observed on the pure Pt catalyst. Following the rapid onset of CO_{ads} oxidation peaking at 0.48 V_{RHE} we

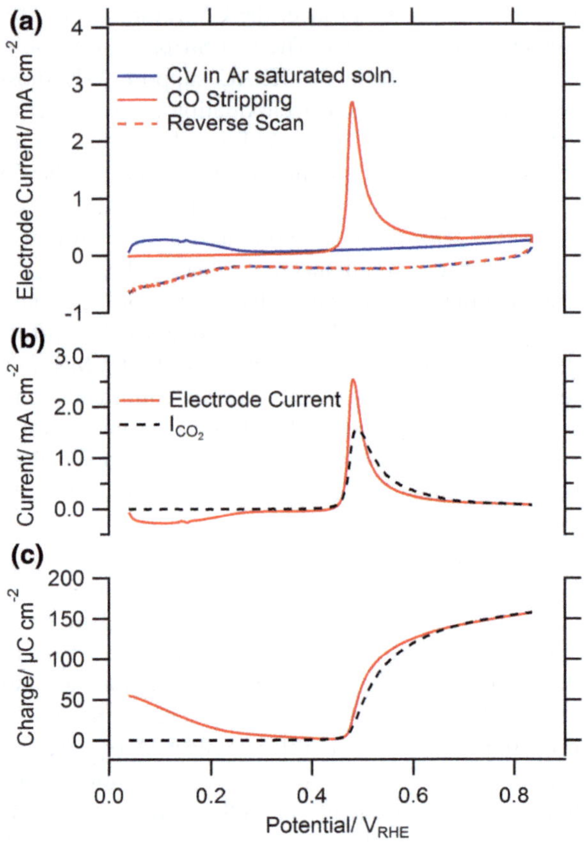

Fig. 5.3 A DEMS CO-Stripping measurement of Tanaka PtRu HSAC supported catalyst. **a** CO-Stripping and background CV in Argon saturated solution, **b** corresponding background subtracted CO-Stripping LSV and I_{CO2}, and **c** charge, performed at 10 mV s^{-1} in 0.5 mol dm^{-3} H_2SO_4. The charge arising from the background subtracted CO-Stripping measurement can be almost exclusively attributed to the oxidation of the saturated CO_{ads} layer

observe an initially rapid fall in current which then begins to slow at 0.52 V_{RHE} broadening the CO_{ads} oxidation peak across a full 400 mV range. This is significantly broader than the symmetric CO-Stripping peak observed on the pure Pt catalyst which possessed a peak width of only 120 mV. Once again, the CV in Argon saturated electrolyte can be subtracted to yield the CO-Stripping and H_{upd} charge and the I_{CO2}observed does not perfectly overlap with the faradaic electrode current. The oxidation of a saturated CO_{ads} layer on the PtRu catalyst at 0.5 V_{RHE} is significantly earlier than on the Pt 5 nm catalyst, and is promisingly within DMFC anode relevant potentials <0.5 V_{RHE}.

The ECSA measurement values and DEMS cell collection efficiency obtained from CO-stripping experiments averaged over ten HSAC –Pt and –PtRu catalyst samples are summarised in Table 5.1.

Here we see that the mass normalised CO-Stripping charges and ECSA are reasonably comparable for both the HSAC –Pt 5 nm and –PtRu catalyst samples. An important observation is that the DEMS cell collection efficiencies for both the Pt and PtRu catalysts were observed to be identical. This in itself might not seem

Table 5.1 Summary of the CO-stripping measurement details and ECSA values of the Tanaka Pt 5 nm and PtRu HSAC supported catalysts

	Pt 5 nm	PtRu
H_{upd} charge/ mC cm^{-2}	6.2 ± 6	4.7 ± 10
CO-stripping charge/ mC cm^{-2}	12.6 ± 10	14.5 ± 14
H_{upd} : CO-stripping charge ratio	1:2	1: ~ 3
Roughness factor/ 50 μg_{metal}	31.5	37.18
ECSA/ m^2 $g_{metal}{}^{-1}$	63	74
DEMS cell collection efficiency/ N_{CO2}	0.3 ± 0.03	0.3 ± 0.03

The ECSA of the PtRu catalysts is comparable to the Pt 5 nm suggesting similar particle sizes, while the DEMS cell collection efficiencies are identical

overly significant, however, it has been noted in previous DEMS studies on PtRu based catalysts that the observed faradaic CO-Stripping charge contains significant current arising from a 'large and undefined capacitive contribution due to the large double layer current' [8, 13] and therefore the ECSA can be overestimate by up to 50 %, but is be avoided using DEMS. The connotations of this claim is that investigations of PtRu using standard RDE methodologies overestimate, by an unknown factor, the ECSA of the PtRu samples (when using CO-Stripping) rendering any subsequent comparison of ECSA normalised activities systematically erroneous. Whilst the investigation into this claim was not the focus of this study, in principle the calculated DEMS collection efficiency, *N*, obtained during the CO-Stripping measurement should be statistically lower for the PtRu catalysts, if indeed the CO-Stripping charge is systematically overestimated because of undefined capacitive current contributions. For this PtRu catalyst, however, this was clearly not the case, although there is a discrepancy in the ratio of the H_{upd} and CO-Stripping charges should be 1:2 (as is the case on Pt electrodes) but is 1:~3 for the PtRu catalyst.

A crucial difference between the observations in the CO-Stripping measurements performed in this study in comparison to the observations in previous studies [8, 13] could be attributed to data treatment rather than the measurement methodology itself. In this study the background CV in Argon saturated solution was subtracted prior to the integration of the CO-Stripping peak for both HSAC supported –Pt and –PtRu catalysts. The background capacitive contributions from the Pt/PtRu surface, the WE substrate and HSAC support are consequently removed and there is no longer an issue regarding the possible over-estimation of the CO-Stripping charge. For PtRu catalysts this background subtraction is perhaps even more important given the greater width of the characteristic CO-Stripping peak (~350 mV), typically ~3–4 times broader than that on Pt (150 mV) with inherently more capacitive contributions. In the previous studies, however, it was not clear if the background CV was subtracted prior to integration, and because it was not explicitly stated then it can only be assumed that this was not the case [8, 13]. The limitation of the findings in the study here, is that it only compares a single PtRu catalyst with Pt, and therefore the observation warrants additional investigation on a number of PtRu catalyst samples in order to clarify this further. A consistent methodology, however, is crucial for electrocatalyst development if we consider the importance of the ECSA on electrocatalytic reaction rates and the comparison of specific (surface area normalised) activities between different electrode materials and research groups.

5.3.2 Cyclic Voltammetry

In this study, the MOR on the HSAC Pt catalyst was first probed using CV. The steady-state CV and I_{CO2} MSCV for the MOR, along with the corresponding calculated conversion of methanol to CO_2 on a HSAC supported Pt 5 nm catalyst is shown in Fig. 5.4.

The CV closely resembles the characteristic form of the potential dependence of the MOR reaction reported previously [9, 33, 34], with the onset of the MOR in the positive potential sweep occurring at ~0.55 V_{RHE} rising rapidly until 0.70 V_{RHE} before slowing and peaking at 0.82 V_{RHE}. At higher potentials the MOR current then decays significantly reaching a minimum at 1.05 V_{RHE}. In the reverse potential sweep, the CV resembles the upward sweep but with a prominent hysteresis, whose peak shifts downwards to 0.65 V_{RHE}. The peak current on the negative scan is furthermore 1.2 mA cm^{-2} higher compared than the peak current in the positive going scan. The relative peak heights of the CO_2 observed in the MSCV during the forward and reverse potential sweep is, however, inverse to the CV and is best expressed in terms of conversion. During the forward potential scan, the conversion of the MOR to CO_2 increases with increasing potential, beginning at ~0.1 and rising to ~0.4 at 0.82 V_{RHE} before decaying to ~0.15 as the potential is increased further. In the reverse potential scan, the conversion of methanol to CO_2 rises only slightly, remaining steady at around 0.25 until falling sharply as the MOR halts. At all electrode potentials, it is clear that we do not come close to the complete conversion of methanol to CO_2 and that the majority of the oxidation of methanol on Pt is only the partial MOR. The highest conversion, however, corresponds to the peak MOR rate indicating a close link between the MOR rate and the conversion of methanol to CO_2.

Although in this DEMS study we were able to present the potential dependence of the conversion of methanol to CO_2, this has not to the authors knowledge been previously presented using CV in the DEMS scientific literature. The conversion of ~0.25 averaged over the whole MOR CV in 1.0 mol dm^{-3} methanol is nonetheless comparable to literature values of 0.28 for polycrystalline Pt [15], but is somewhat lower than the 0.65 reported previously on a HSAC support pure Pt [9] in 0.1 mol dm^{-3} methanol. Although still far less than 1.0, the averaged conversion values observed using DEMS have been shown to strongly depend on the experimental parameters [11, 15, 31].

The steady CV and MSCV of I_{CO2} for the MOR, along with the corresponding calculated conversion of the MOR to CO_2 over a HSAC supported PtRu catalyst is shown in Fig. 5.5.

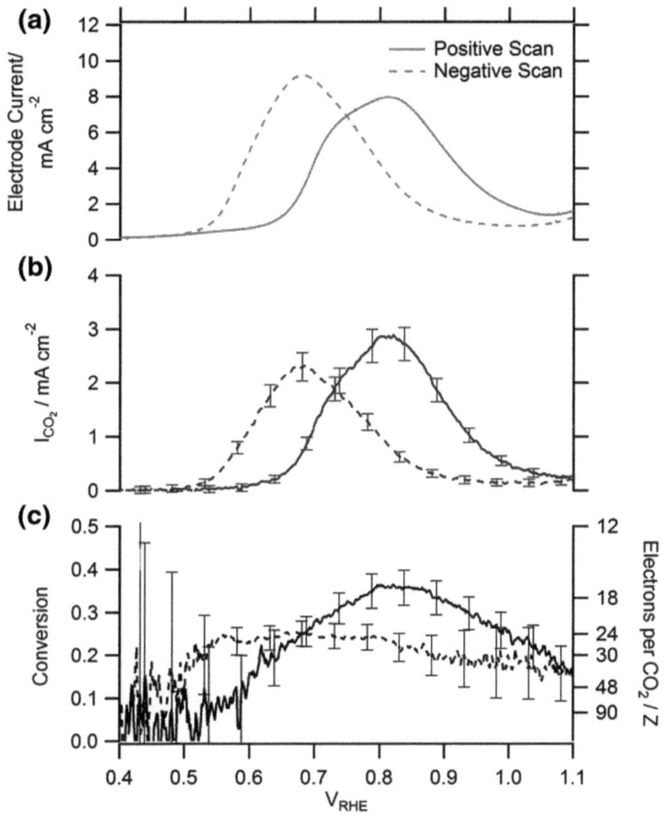

Fig. 5.4 DEMS measurement of the MOR on HSAC supported Pt 5 nm catalyst. CV of 1.0 mol dm^{-3} methanol oxidation on HSAC supported Tanaka Pt 5 nm catalyst (**a**), corresponding MSCV for I_{CO2} (**b**), and resulting calculated conversion of methanol to CO2 (**c**), at 10 mV s^{-1} in 0.5 mol dm^{-3} H_2SO_4 and an electrolyte flow rate of 5 μl s^{-1}. The conversion of the MOR to CO_2 is far from complete, and is furthermore potential dependent.

The CV closely resembles the characteristic form of the potential dependence of the MOR reaction observed in previous studies [8, 33], with the onset of the MOR in the positive potential sweep occurring at 0.33 V_{RHE} rising rapidly until ~0.60 V_{RHE} before slowing and peaking at ~0.70 V_{RHE}. At higher potentials the MOR current begins to decay, however, the CV in this case is limited to an upper potential of 0.85 V_{RHE} to prevent the preferential dissolution of Ru at more positive potentials [20]. In the reverse potential sweep, the MOR current is ~50 % lower with little or no hysteresis, and no clear peak in the observed current or MSCV. Significantly, the corresponding potential dependent conversion of the MOR–CO_2 on the PtRu catalyst is strikingly different to the potential dependent conversion on the pure Pt catalyst. In the case of the PtRu catalyst, using DEMS we observe that the conversion actually decreases from >0.7 at 0.45 V_{RHE} to

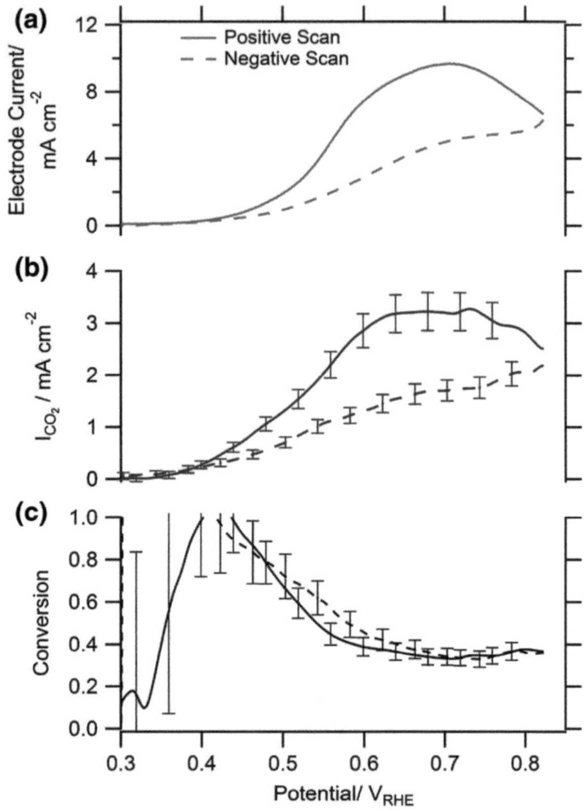

Fig. 5.5 DEMS measurement of the MOR on Tanaka HSAC supported PtRu catalyst. CV of 1.0 mol dm^{-3} methanol oxidation on HSAC supported Tanaka PtRu catalyst (**a**), corresponding MSCV for I_{CO2} (**b**), and resulting calculated conversion of methanol to CO_2 (**c**) at 10 mV s^{-1} in 0.5 mol dm^{-3} H_2SO_4 and an electrolyte flow rate of 5 μl s^{-1}. The conversion of the MOR to CO_2 on PtRu is potential dependent, decreasing with increasing MOR rate

~0.35 at 0.7 V_{RHE} despite the increase in electrode potential and MOR rate. This trend is furthermore very similar during both the positive and negative potential sweep, suggesting little change in the MOR mechanism, despite the significantly smaller electrode current.

Similarly to the Pt catalyst, the potential dependence of the conversion of the MOR to CO_2during a CV on the MOR on a HSAC supported PtRu surface has not been reported previously in the DEMS scientific literature. The average conversion of the MOR to CO_2 observed in this experiment of ~0.5 in 1.0 mol dm^{-3} methanol is lower than the average value of ~0.9 (expressed as slightly above 6 electrons per CO_2) in 0.1 mol dm^{-3} given previously [8]. The higher averaged conversion on the PtRu ~0.5 catalyst in comparison to ~0.25 on the Pt 5 nm catalyst is meanwhile in agreement with previous comparisons [31]; however, again the exact values have been shown to be dependent on the experimental parameters [11, 15, 31]. The trends for higher methanol conversion to CO_2 with slower electrolyte flow rates [15], lower methanol concentrations [15, 31] and higher catalyst loadings [11] observed in these previous studies, are strongly indicative of a MOR mechanism which involves more than a single adsorption step i.e. the MOR largely proceeds via the in-direct pathway involving desorbed intermediates.

In the interpretation of the contrasting potential dependant conversion trends observed during the CV, we must be aware that the I_{CO2} and therefore conversion is to some extent overestimated during the onset of the MOR in the positive potential sweep. At potentials below the onset of the MOR <0.35 V_{RHE} the electrode surface is already covered in partial MOR products arising from the dissociative adsorption of methanol at potentials as low as 0.1 V_{RHE} (not shown in these measurements but examples may be found in Ref. [3, 4]) which have been deduced to largely consist of CO_{ads} species [8, 9]. The oxidation of these species would therefore yield fewer than the six electrons per CO_2 assumed for the oxidation of methanol in the calculation of the I_{CO2} leading to the overestimation of the conversion until all the previously adsorbed partial MOR products have been oxidised. The coverage of these CO_{ads} species has been observed to be ~0.75 for pure Pt catalysts [9] and <0.4 for PtRu [8] which may suggest Pt is most overestimated. Assuming that the partially oxidised MOR products oxidise at the onset of the MOR LSV, we can deduce that the charge arising from the oxidation of the previously adsorbed partial MOR products has dissipated prior to ~0.60 and ~0.43 V_{RHE} for the Pt and PtRu catalyst respectively. The observed potential dependent trends are therefore still valid; however, we should be aware that the conversion is to some extent overestimated during the course of the CV.

Before rationalising the observed potential dependence, it is crucial to first examine the relationship between the MOR and oxidation of CO_{ads} (which has been well documented in the scientific literature) by comparing the LSV of the MOR and the CO-Stripping measurement, and to later confirm the potential dependent conversion trend of the MOR using sampled current chronoamperometry, which is then extended to three-dimensional voltammetry.

5.3.2.1 Relationship Between Methanol and CO_{ads} Oxidation

The relationship between the CO_{ads} oxidation and the MOR on Pt HSAC supported catalyst was examined by comparing the positive MOR LSV and I_{CO2} mass spectrometric LSV (MSLSV) with the electrode current for oxidation of a CO_{ads} saturated layer, as shown in Fig. 5.6a.

In the CV and MSLSV (I_{CO2}) we observe that the MOR peak current potential and CO-Stripping peak current closely coincide. This correlation is more precisely observed in the differential voltammogram in Fig. 5.6b, whereby the MOR and CO-Stripping LSV peak maximums (where $dI/dV = 0$) coincide at precisely 0.815 V_{RHE}. An additional feature of the differential voltammogram, are the peaks that indicate the potential at which the current rises fastest during the LSV. This peak MOR potential in the positive going potential scan is the same for both the electrode current, and I_{CO2} which is found at 0.695 V_{RHE}, which interestingly coincides closely with the onset potential for the oxidation of a saturated monolayer of CO_{ads} (seen more precisely in the differential plot). Nonetheless, there is clearly a close relationship between the ability of the Pt catalyst to oxidise a saturated CO_{ads} layer and the MOR rate on the HSAC Pt catalysts.

Fig. 5.6 A comparison of the MOR and CO-Stripping LSV of Pt 5 nm HSAC supported catalyst. LSVs for electrode current and I_{CO2} (**a**) and corresponding differential voltammogram (**b**) of HSAC supported Tanaka Pt 5 nm catalyst in 0.5 mol dm^{-3} H_2SO_4 supporting electrolyte during 1.0 mol dm^{-3} methanol oxidation and CO-stripping, with a potential scan rate of 10 mV s^{-1} and electrolyte flow rate of 5 μl s^{-1}. The key features of the LSV for the MOR on the Pt catalyst are closely related to the LSV of the oxidation of a saturated CO_{ads} layer

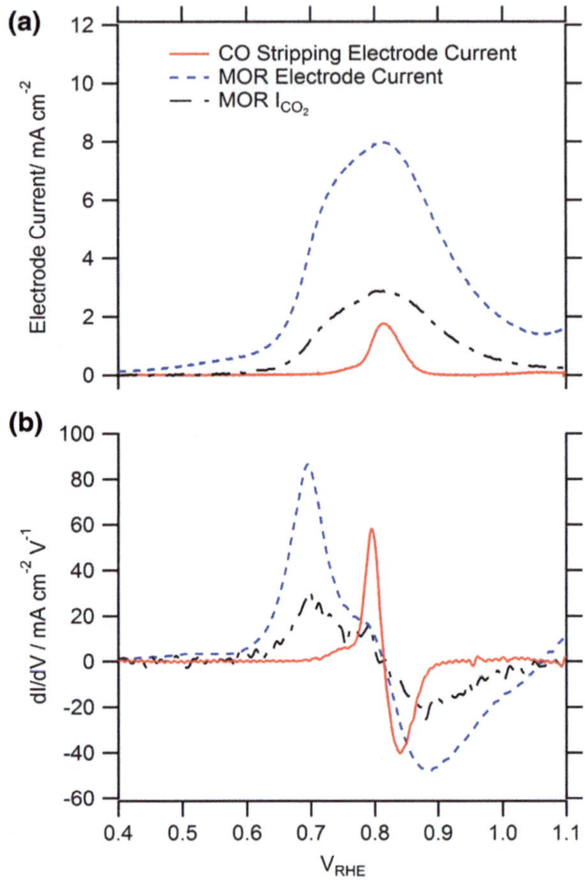

If we compare the MOR LSV and I_{CO2} MSLSV to the oxidation of a saturated CO_{ads} layer on the PtRu HSAC supported catalyst in Fig. 5.7, we observe fewer similarities between the features of the MOR CV and CO-Stripping LSV.

The oxidation of a saturated layer of CO_{ads} on the PtRu catalyst occurs at ~0.4 V_{RHE}, peaking at 0.46 V_{RHE}, 100 mV lower in potential than the peak MOR current at 0.56 V_{RHE} observed in the LSV and MSLSV. The MSLSV for I_{CO2} however, possesses a much broader peak that remains fairly constant between 0.64 and 0.74 V_{RHE}. The highest potential dependence on the current of the MOR also follows the CO-stripping onset by 200 mV. Although the earlier onset of the MOR at 0.33 V_{RHE} on the PtRu catalysts compared with 0.55 V_{RHE} on the Pt catalyst of the MOR can be explained by the co-catalytic mechanism of the PtRu surface and the ability of Ru-OH_{ads} species to oxidise poisoning Pt-CO_{ads} which prevent the initial dissociative adsorption of methanol; the MOR electrode current (and therefore reaction rate) does not appear to be so closely related to the ability for the HSAC supported PtRu catalyst to oxidise CO_{ads} species.

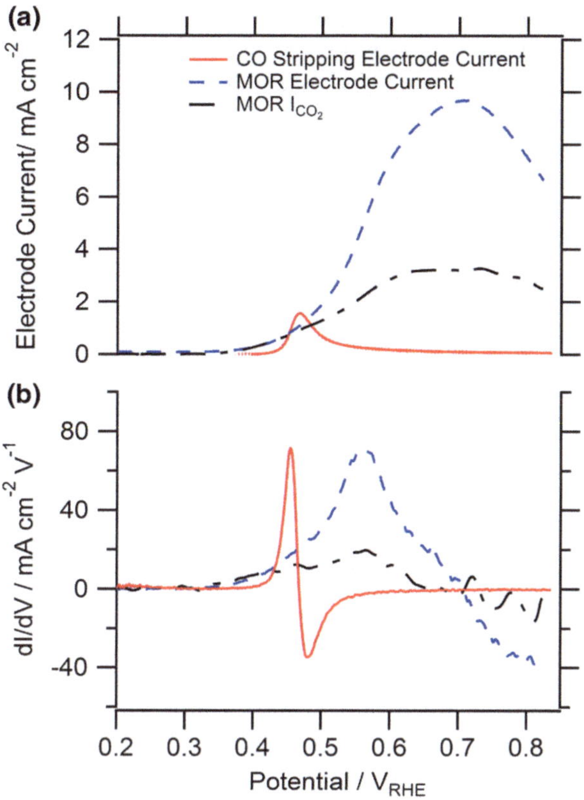

Fig. 5.7 Comparison of the MOR and the CO-stripping of the Tanaka HSAC support PtRu catalyst. LSV and MSLSV (**a**) and corresponding differential voltammogram (**b**) of HSAC supported Tanaka PtRu catalyst in 0.5 mol dm^{-3} H_2SO_4 supporting electrolyte during 0.1 mol dm^{-3} methanol oxidation and CO-stripping. The features of the CV for the MOR on the PtRu catalyst do not correlate closely with the oxidation of a saturated CO_{ads} layer

It is clear from these comparisons that the relationship between the MOR and the oxidation of a saturated CO_{ads} layer is quite different for the Pt in comparison to the PtRu catalyst. Whereas the features of the LSV of the MOR on the Pt catalysts correspond meticulously with the features of the CO-Stripping measurement indicating a close relationship between the two, the features of the MOR and CO-Stripping LSVs of PtRu catalyst are somewhat different. All measurement observations so far have, however, been performed using potentiodynamic electroanalytical techniques, and it is therefore not clear at this point if these trends are reflected in potentiostatic steady-state conditions. Consequently, the observations made of the MOR using CV are now compared with sampled current chronoamperometry in the following section.

5.3.3 Chronoamperometry

Chronoamperometry is a powerful electrochemical technique for studying electrochemical systems by monitoring the current as a function of time at a certain

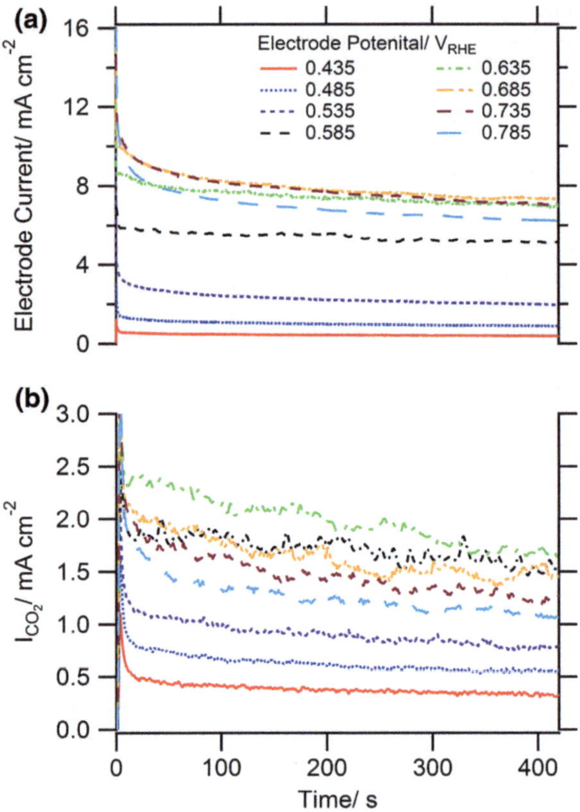

Fig. 5.8 A plot of multiple chronoamperometric (current vs. time) measurements for the MOR at a number of electrode potentials (**a**) and corresponding I_{CO2} response (**b**). Each chronoamperometric measurement was performed by stepping the electrode potential from 0.05 V_{RHE} to the desired electrode potential, which was progressively increased in 50 V steps from 0.435 to 0.785 V_{RHE}

fixed potential, and has been used previously in kinetic studies on the MOR on Pt and PtRu based catalysts [35–41]. Example chronoamperometric curves for the MOR, and corresponding I_{CO2} on the PtRu catalyst performed at numerous electrode potentials are given in Fig. 5.8.

In the current response to the potential step there is the typical initial capacitive current spike, along with MOR current which quickly decays to a steadier faradaic MOR current after ~1 s, the exact value depending on the size of the potential step. The MOR current then decays gradually over time at all potentials during the course of the chronoamperometric measurement, a trend that is also observed in the corresponding I_{CO2}. The presentation and discussion of chronoamperometric data at multiple potentials is, however, somewhat difficult to visualise using such current vs. time plots and consequently the current from a multiple of chronoamperometric measurements at various potentials are often sampled at a certain time and plotted as function of electrode potential—a technique that is often referred to as sampled current chronoamperometry. This aids in expressing chronoamperometric data as a function of potential which in turn allows the comparison of chronoamperometric measurements with CVs.

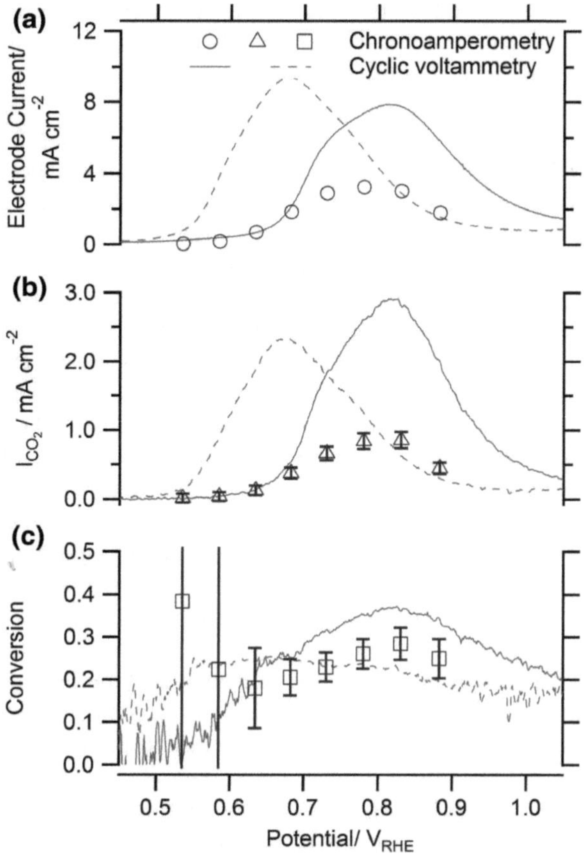

Fig. 5.9 Comparison of sampled current chronoamperometry with CV for the (MOR) on Tanaka HSAC supported 5 nm Pt catalyst. Sampled current chronoamperometry of 1.0 mol dm^{-3} methanol oxidation (**a**), corresponding I_{CO2} (**b**), and resulting calculated conversion of methanol to CO_2 (**c**), sampled at 420 s at an electrolyte flow rate of 5 μl s^{-1}. The CV from Fig. 5.4 is also shown (in *grey*) for direct comparison. The sampled current chronoamperometric measurements confirm the trends observed in the CV

The sampled current chronoamperometric measurements sampled at 420 s for the MOR on the Pt 5 nm catalyst, corresponding I_{CO2} and calculated conversion of methanol to CO_2 is given in Fig. 5.9. Note that the traces of the CV are also indicated on the plot to aid in the comparison of techniques.

In general, the trends observed in the chronoamperometric data are in agreement with those observed in the CV. The steady-state electrode and I_{CO2} correspond reasonably well with the CV at potentials <0.75 V_{RHE}, where there is relatively little MOR, however, the steady-state faradaic current deviates to significantly lower values at higher electrode potentials indicating that the MOR rate has decayed significantly overtime on the Pt 5 nm catalyst. The peak electrode current potential also appears to have shifted to lower potentials compared to the CV; however, the peak mass ion current still corresponds well with I_{CO2}. Furthermore, the observed conversion during the potentiostatic measurements on the MOR to CO_2, is lower ($\sim$0.1) than suggested by CV, although the trend for greater conversion at the peak MOR current potential remains the same.

The trend of increasing conversion with MOR rate is in agreement with a previous DEMS study which observed a slight increase in conversion of the MOR to CO_2 based on two chronoamperometric measurements at 0.65 and 0.75 V_{RHE} [14]. This trend also corresponds to the (ex-situ) measurement of formaldehyde yields using fluorescence spectroscopy following chronoamperometric measurements of the MOR on polycrystalline Pt [26]. Here, formaldehyde yields decreased with increasing MOR rates and potentials ranging from 40 to 5 % yields between 0.25 and 0.5 $V_{Ag/AgCl}$ (ca. 0.5–0.75 V_{RHE}) respectively. The conversion of the MOR to CO_2 on a pure Pt electrode using chronoamperometry has also been previously reported to be rather high ~0.85 at 0.55 V_{RHE} using SDEMS during a potentiostatic measurement sampled at 300 s [10]. Unfortunately, it is not possible to extract a comparable value from the data in Fig. 5.9 because the very small electrode and I_{CO2} currents observed at these potentials induce significant errors in the resulting calculated conversion. Furthermore, there is no trend in these measurement data to suggest that the conversion is increasing towards lower electrode potentials and the high conversion at this potential claimed previously can perhaps be treated with some caution. Overall, it is clear that the study of the MOR activity using CV is not representative of the time resolved steady-state MOR electrode current of the HSAC supported Pt catalyst obtained using chronoamperometry, with significantly lower MOR electrode current and conversion to CO_2 observed.

The sampled current chronoamperometric measurements taken at 420 s for the MOR on the PtRu catalyst, corresponding I_{CO2} and calculated conversion to CO_2 is given in Fig. 5.10.

It is clear in this plot that the potentiostatic measurements follow the same trend as the CV; however, the steady-state faradaic currents are all ~20 % lower across the whole potential range in contrast to the Pt 5 nm catalyst which was ~60 % lower. The chronoamperometric data also confirms the trend for decreasing conversion with higher electrode potentials on the PtRu catalyst; however, the steady-state conversions are ~15 % lower in conversion across the whole potential range compared to the CV. The conversion of the MOR–CO_2 at 0.55 V_{RHE} of 0.3 is comparable to the conversions of 0.2–0.3 observed using DEMS on a PtRu catalysts with a composition of 20–40 % wt. Ru [10]. The potential dependence of the MOR on the PtRu catalyst meanwhile is in agreement with the observations hinted at in the chronoamperometric measurements in Ref. [14]; although the data presented in this study is without question more conclusive. The generally higher conversion observed for the MOR on the PtRu catalyst compared to Pt is also in agreement with previous observations [15]. Measurements of formaldehyde yields using fluorescence spectroscopy following chronoamperometry, however, did not follow this trend, although very low yields (<3 %) for formaldehyde were observed on HSAC supported PtRu (1:1) catalysts [27].

Overall, the sampled current chronoamperometric measurements confirm the potential dependence of the MOR, however, both the electrode current and conversion of methanol to CO_2 are lower under steady-state conditions. Part of the loss of MOR electrode current may be attributed to the adsorption of impurities as assumed in previous studies [6, 35, 42], although in the case of industrial electrocatalysts with large ECSA, the influence of adsorbing impurities is not

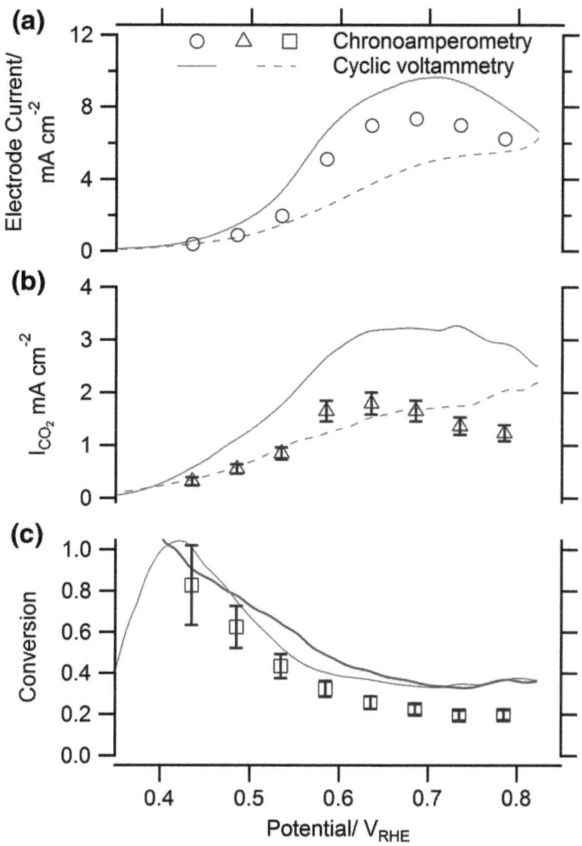

Fig. 5.10 Comparison of the sampled current chronoamperometric measurement with CV for the MOR on Tanaka HSAC supported PtRu catalyst. Sample-current chronoamperometry of 1.0 mol dm^{-3} methanol oxidation on HSAC supported Tanaka PtRu catalyst (**a**), corresponding I_{CO2} (**b**), and resulting calculated conversion of methanol to CO_2 (**c**), sampled at 420 s, with an electrolyte flow rate of 5 μl s^{-1}. The CV from Fig. 5.5 is shown for comparison. The sampled current chronoamperometric measurements confirm the trends observed in the CV

comparable to model electrocatalysts which possess considerably lower ECSA (and are thus more sensitive to electrolyte impurities).

Although the chronoamperometric measurements are very useful in assessing the steady-state MOR currents (which is more crucial to DMFC development) it is unfortunate that in performing these measurements and creating plots of this type, much of the experimental data is actually discarded. For example, it is not easy to convey how the current reached a certain current value over time, as a function of potential in such a two dimensional format. Another possibility is to plot the three measurement variables (potential, current and time) in the form of a three-dimensional voltammogram, which to the author's knowledge has not been previously applied to the study of the MOR. Such plots can aid in visualising the experimental data and by providing a new perspective could allow further interpretation of the time dependent behaviour of the MOR system to be made.

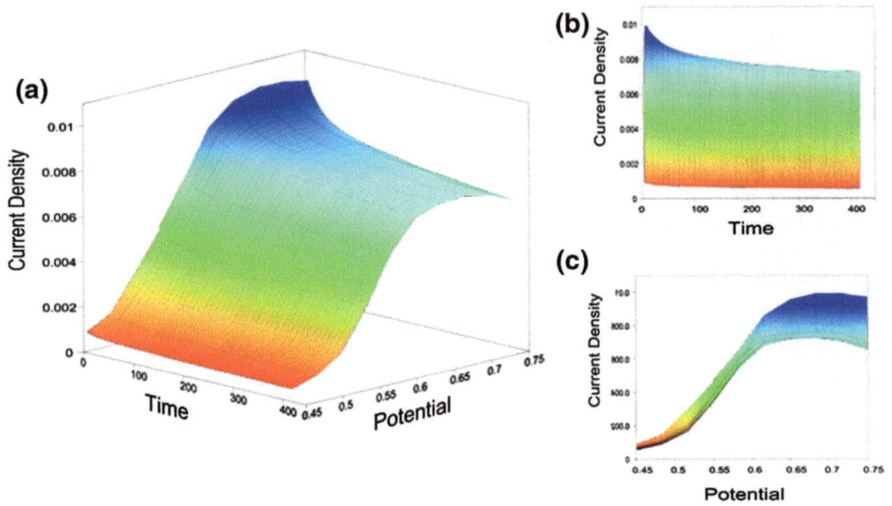

Fig. 5.11 Three-dimensional voltammogram of 1.0 mol dm^{-3} methanol oxidation on HSAC supported Tanaka PtRu catalyst.The 3-dimensional plot encapsulates the current (A cm^{-2}), potential (V_{RHE}) and time (s) relationship of the MOR within a single plot

5.3.4 Three-Dimensional Voltammetry

The three-dimensional voltammogram that is depicted in Fig. 5.11a was constructed to encapsulate all the data contained within a series of chronoamperometry measurements, rather than discarding much of the data as is the case in sampled current chronoamperometry. This allows the visualisation of the current, time and potential relationship within a single plot, which could allow further interpretation of the electrochemical data and reaction mechanisms.

By rotating the orientation of the three-dimensional plot we can also observe the current versus time Fig. 5.11b, and the current versus potential relationships Fig. 5.11c. Although this three-dimensional voltammogram aids in visualising the reaction system, it is impossible to deduce specific information from the plot in its current form. In a previous study on an unrelated reaction system that also produced these three-dimensional plots, much of the data was after all presented and analysed in the form used for sampled current chronoamperometry [43] although the author compared different points in time. A potentially valuable alternative representation of the three-dimensional data which could offer a new perspective and interpretation of a reaction system (which to the author's knowledge has not been used previously to describe an electrochemical half-cell reaction) is the contour surface plot.

5.3.4.1 Contour Surface Plot

The contour surface plot represents three-dimensional data by plotting slices of the current data in a two dimensional format, allowing the mapping of the current,

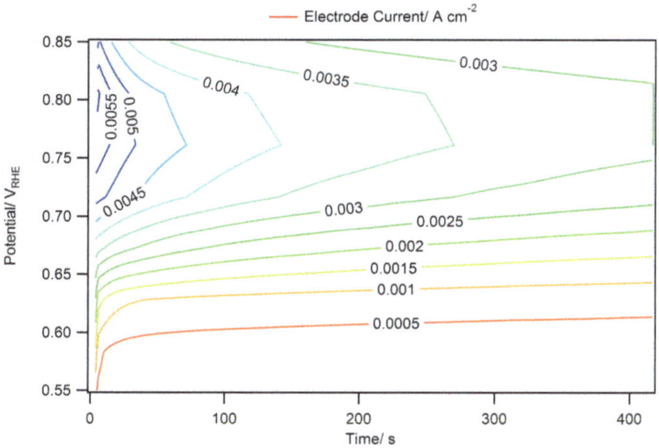

Fig. 5.12 Contour plot of 1.0 mol dm^{-3} methanol oxidation on HSAC supported Pt catalyst. Electrolyte flow rate of 5 μl s^{-1} in 0.5 mol dm^{-3} H_2SO_4, contour units in A cm^{-2}. The contour plots of the three-dimensional plot provide a new perspective on the MOR.

potential and time relationship of the reaction system in a single interpretable format. The current contours describing the MOR on the HSAC supported Pt catalyst is presented in Fig. 5.12.

The fan shaped form of the current contours identifies the potential region of the MOR system on the Pt catalyst, and its change in potential over time as the current decreases. This brings greater focus on the potential region between ~0.58 and 0.68 V_{RHE} where the MOR current is relatively stable over the course of the experiment, in comparison to potentials >0.7 V_{RHE} the MOR electrode current decays more rapidly with time, whilst furthermore highlighting the lack of MOR current <0.55 V_{RHE}. The form of the curves furthermore resemble those of chronopotentiometric measurements (a.k.a. galvanostatic polarisation curves) such as those given in Ref. [44].

By extending this technique to incorporate DEMS data for quantitative measurement of the CO_2, the conversion of the MOR over the whole current, potential and time relationship may also be expressed in the form of the contour plot. In order to compare the relationship between the conversion of the MOR and the faradaic electrode current, the conversion contour plot is overlaid onto the faradaic electrode current in Fig. 5.13.

Here we can observe that the form of the contours of the conversion correlate closely with the contours of the electrode current, particularly in the lower potential regions <0.7 V_{RHE}. This is able to illustrate the region in which there is the crucial link between the observed MOR current and the conversion of the MOR–CO_2 (and hence CO_{ads} oxidation) over the whole MOR system within the time frame of the experiment, an observation suggested previously by the CV and chronoamperometric measurements in this study but represented fully in the contour plot.

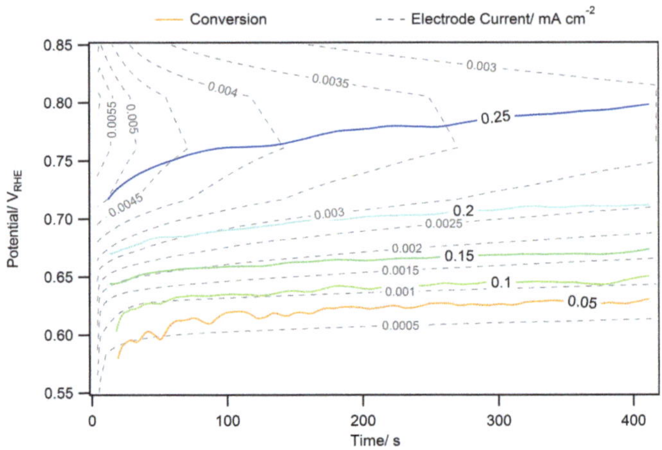

Fig. 5.13 Contour plot of the conversion (*coloured*) of 1.0 mol dm^{-3} methanol oxidation to CO_2 on HSAC supported Pt 5 nm catalyst. Electrolyte flow rate of 5 μl s^{-1} in 0.5 mol dm^{-3} H_2SO_4, and contours (*greyed*) of the faradaic electrode current given in Fig. 5.12. The contours of the conversion of the MOR to CO_2 closely correlate with the contours of the faradaic electrode current

The corresponding contour plots representing the electrode current and conversion of the MOR to CO_2 on the PtRu catalyst is presented in Fig. 5.14.

While the fan shaped contours of the MOR electrode current observed on the PtRu catalyst in Fig. 5.14a are analogous to those observed on Pt, the contours of the conversion of the MOR to CO_2 on the PtRu depicted in Fig. 5.14b do not resemble the shape and form of the electrode current, as was the case on the pure Pt catalyst. The conversion clearly decreases with time, particularly as the potential is increased >0.55 V_{RHE}; however, despite this decrease the electrode current of the MOR is maintained. This would suggest that the oxidation of CO_{ads} species is less crucial of the MOR rate on the PtRu catalyst, and that the MOR mechanism is not only changing as a function of potential, but also as a function of time during the course of the chronoamperometric measurement. Although the rationalisation of this observation is not entirely clear, this change may represent a gradual alteration in the surface morphology of the PtRu catalyst surface which begins to occur at potentials as low as 0.55 V_{RHE}, an observation that cannot be deduced from the electrode current alone. For example, the Pt character of PtRu electrodes has been shown to gradually increase with potential cycling up to 0.7 V_{RHE} [45] whilst the preferential dissolution of Ru has also been observed in DMFC anodes [46]. Essentially, by measuring the MOR at potentials using chronoamperometry above 0.55 V_{RHE} we perhaps irreversibly change the surface morphology of the PtRu catalyst.

5.3.4.2 Potential Uses of Contour Surface Plots in Electrochemistry

The contour plots presented in this section offer a new perspective on the MOR, allowing the current, potential and time relationship of the reaction system to be

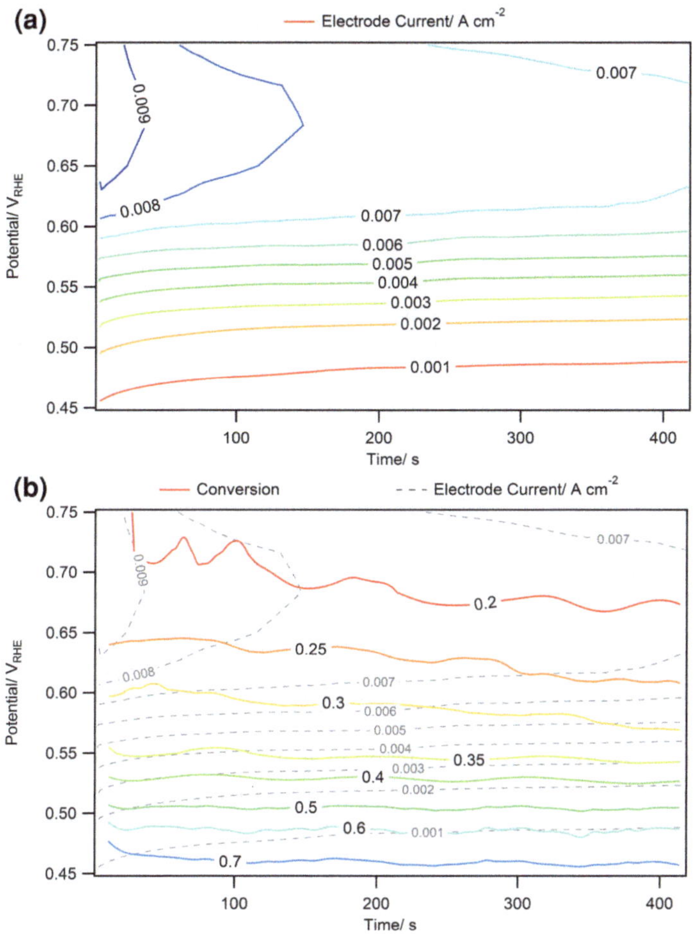

Fig. 5.14 Contour plot of the faradaic electrode current (**a**) and conversion (**b**) of 1.0 mol dm^{-3} methanol oxidation to CO_2 on the HSAC supported Tanaka PtRu catalyst.Electrolyte flow rate of 5 µl s^{-1} in 0.5 mol dm^{-3} H_2SO_4. The contours of the faradaic electrode current are also displayed in *grey* in (**b**) for comparison. The contours of the conversion of the MOR to CO_2 do not correlate with the contours of the faradaic electrode current

expressed in a single interpretable format. Besides being a tool to aid in the visualisation of the MOR system (or indeed any electrochemical reaction system), it can also aid in new interpretation, identify regions of interest or to confirm findings over the whole reaction system during the course of a half-cell measurement. Combined with techniques such as DEMS, the plots can furthermore be used to map the conversion or product distributions of the reaction system to at least contribute towards our understanding of the MOR system in electrochemical half-cell. Moreover, the three-dimensional plots are conceivably very useful in aiding in the planning of chronopotentiometric measurements which are often notoriously difficult to perform in fundamental studies. In such measurement, meticulous knowledge of the current, potential and time relationship of a reaction system is required to ensure that the potential of the electrode (whilst not under direct control) is coerced within an acceptable or desired potential region. This is crucial in electrocatalysis because the structure of the electrode is sensitively dependent on the potential (often irreversibly so) and avoiding high potentials that can modify or destroy pristine electrode samples is essential. This is of course assuming that the electrochemistry of an electrode reaction is not influenced by the electroanalytical technique itself. For instance, the oscillation of the electrode potential that can be observed in some reaction systems (such as formic acid oxidation) using chronopotentiometric measurements cannot be resolved using chronoamperometry.

Although three-dimensional plots have been suggested a number of times, and demonstrated for a limited number of reaction systems [43], there are rather few or no other examples of their use in electrochemistry to convey specific information. This is certainly, partly because of the limitations described in Ref. [43] such as the lack of automated setups along with instrumentation restrictions, but there is also the difficulty in expressing the data contained in the three-dimensional plot which, although look nice and are useful for educational purposes, are not particularly interpretable for more quantitative information. Hopefully current contour plots help to convey this information and could be potentially useful in encapsulating and elucidating many electrochemical processes in more than just a purely qualitative way.

Having established an understanding of the MOR system using CV, chronoamperometry and three-dimensional voltammetry, it is now useful to evaluate the kinetics and activity of the Pt and PtRu catalysts towards the MOR.

5.3.5 Tafel Slope

The logarithmic current vs. potential plot (or Tafel plot) and corresponding I_{CO2} for the MOR on the HSAC supported Pt catalyst measured using both CV and sampled current chronoamperometry is presented in Fig. 5.15a and b respectively.

The linear region (or rather Tafel-slope region) observed between 0.65 and 0.7 V_{RHE} for the MOR of 82 mV Dec^{-1} of current on the Pt catalyst during the LSV is mirrored precisely by the Tafel-slope of 79 mV Dec^{-1} for the I_{CO2} between 0.55 and 0.7 V_{RHE} observed using DEMS, and corresponds closely with literature values of 80 mV Dec^{-1} [5, 35]. Although the Tafel behaviour is only observed over less than one decade of current during the LSV, the potential range is extended when using sampled current chronoamperometry (there are for instance no influences from capacitive currents), although a slightly shallower slope of 96 mV Dec^{-1} is observed. The linearity in the LSV between 0.45 and 0.60 V_{RHE} for the electrode current meanwhile has less significance owing to the rather small electrode currents involved, a large proportion of which is likely to be

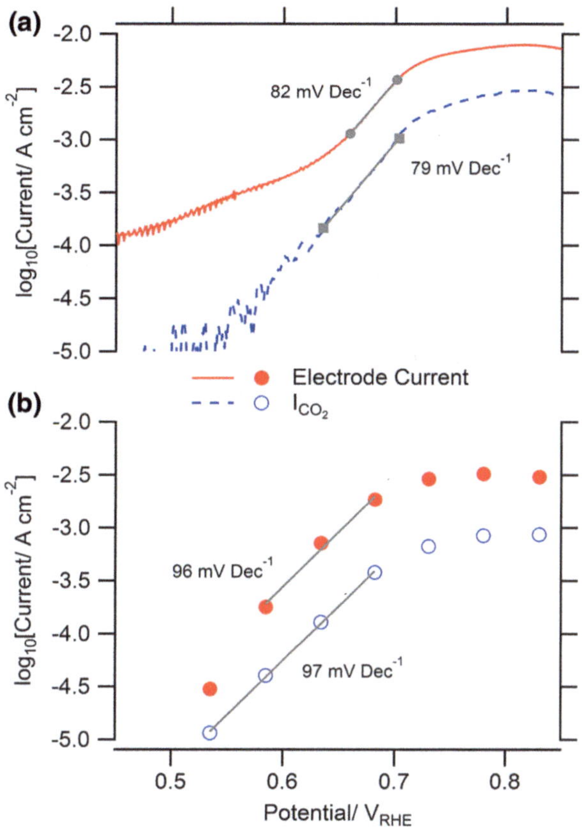

Fig. 5.15 Tafel-slope of LSV for 1.0 mol dm^{-3} methanol oxidation on HSAC supported Tanaka Pt 5 nm catalyst and MSLSV for I_{CO2} at a potential scan rate of 10 mV s^{-1} (**a**) and corresponding chronoamperometric current samples at 420 s (**b**) electrolyte flow rate of 5 μl s^{-1}. The Tafel slope of the faradaic electrode current closely corresponds to that of I_{CO2}

capacitive in nature and are therefore short lived. Nonetheless, the Tafel-slope for I_{CO2} corresponds to the MOR electrode current on the Pt catalyst and identifies the potential region where the current scales exponentially with the electrode potential and is should be the region of interest in the studies of electrode kinetics.

The Tafel-slopes observed for the electrode current, and I_{CO2} for the MOR on the PtRu catalyst during LSV and sampled current chronoamperometry are given in Fig. 5.16.

In the both the LSV and the sampled chronoamperometric measurement of the MOR on the PtRu catalyst, a clear Tafel-slope of 134 and 129 mV Dec^{-1} is observed over more than one decade of current between 0.40 and 0.55 V_{RHE}. This value is within the range of literature values obtained from previous electrochemical half-cell studies (typically RDE) such as 115 mV Dec^{-1} for PtRu/C [42], 100 mV Dec^{-1} on PtRu/Nanotubes [37], 195 mV Dec^{-1} on PtRu/Vulcan [35], 180–195 mV Dec^{-1} on PtRu surface alloys [5] and 115 mV Dec^{-1}in $HClO_4$ [47] and finally ~110 mV Dec^{-1} observed for PtRu/C in a DMFC [48]. The corresponding I_{CO2} signal, however, clearly does not follow the same trend with only slight linearity observed between 0.50 and 0.55 V_{RHE} in the CV with a rather

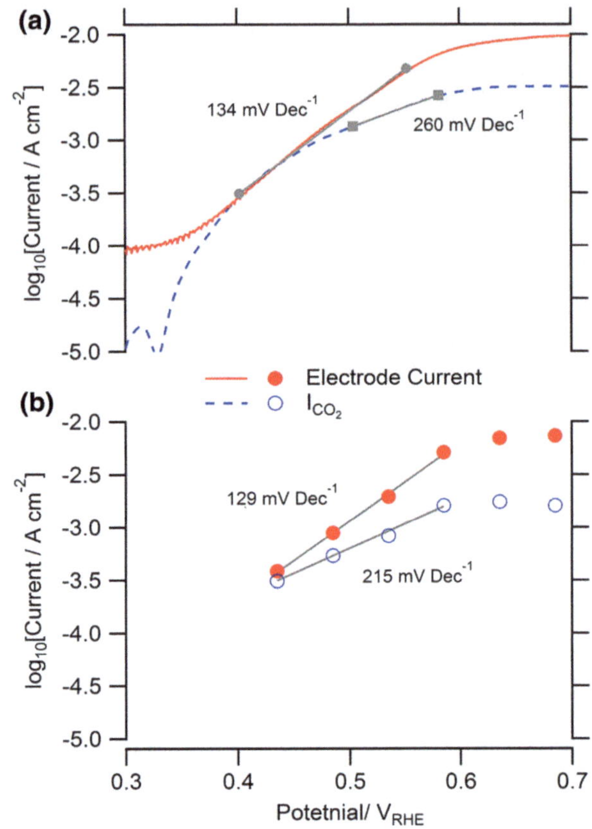

Fig. 5.16 Tafel-slope of the LSV for 1.0 mol dm^{-3} methanol oxidation on HSAC supported Tanaka PtRu catalyst and MSLSV for I_{CO2} at a potential scan rate of 10 mV s^{-1} (**a**) and corresponding chronoamperometric current samples at 420 s (**b**) electrolyte flow rate of 5 µl s^{-1}. The Tafel slope of the faradaic electrode current does not correspond to that of I_{CO2}

shallow linear slope of 260 and 215 mV Dec^{-1}. The curve of the I_{CO2} signal in the LSV below 0.37 V_{RHE} is also comparatively steep, however, in this potential region we cannot distinguish between the MOR and the oxidation of previously adsorbed partial MOR products. Overall, the Tafel slope of the I_{CO2} does not correlate to the MOR electrode current on the PtRu catalyst in contrast to Pt where they were identical, which perhaps reflects the differences in the MOR rate determining step (RDS).

A detailed discussion of the circumstances where mechanistic information can and cannot be extracted from the Tafel-slope is beyond the scope of this DEMS study on the MOR, and the reader is instead directed to Ref. [49] (Chaps. 4 and 12). Briefly, the key points are that the transfer coefficient extracted from the Tafel-slope for a multi-step reaction can allow interpretation of the RDS of the reaction based on the number of electrons involved in the RDS, and the total number of electrons in the reaction by assuming all other reaction steps are comparatively fast. However, the extraction of such mechanistic information is not possible for multi-step reactions that are either bi-functional and/or involve a significant coverage of adsorbed intermediates [49], such as the MOR where rather unusual Tafel-slopes are generally observed. Consequently, the significance of the Tafel-slopes observed for the MOR can and has often only been discussed in a very qualitative manner, whereby it is acknowledged that the Tafel-slope does not carry its usual meaning [5]. This statement is sometimes referenced [37] but rarely elaborated on, with the exception of Ref. [35]. Here it was suggested that it still reflects the nature of the RDS (common to all multi-step reactions) whereby the differences in the Tafel-slope of the MOR on Pt and PtRu were rationalised by proposing that Tafel-slopes of <130 mV Dec^{-1} are typical for surface reaction RDS such as CO_{ads} oxidation on Pt, whereas Tafel-slopes >130 mV Dec^{-1} can be expected from a chemical RDS i.e. the dissociative adsorption of methanol on PtRu. The difference in the MOR RDS on the Pt and PtRu catalyst is also supported by the observation of different activation energies (typically ~60 kJ mol^{-1} on PtRu vs. 17–29 kJ mol^{-1} for Pt), and that the MOR on Pt is essentially zero order at high methanol concentrations >0.5 mol dm^{-3} [5].

The correlation between the Tafel-slope for the MOR electrode current and I_{CO2} on the Pt catalyst (not to mention the close correlation between the MOR and CO_{ads} oxidation) is clearly in accordance with the well accepted conclusion that the CO_{ads} is the RDS. In contrast, the lack of correlation between the Tafel-slope of the MOR electrode current and I_{CO2} on the PtRu catalyst would on the other hand suggest that CO_{ads} oxidation is not the RDS for the MOR on PtRu, assuming all CO_2 is produced via CO_{ads} oxidation. Nonetheless, there is still a clear correlation between the ability of the PtRu catalyst surface to oxidise CO_{ads} and the onset of the MOR. The poisoning CO_{ads} species, therefore, can be concluded to still be inherently responsible for the necessary overpotential for the MOR, even if the oxidation of the CO_{ads} is not the RDS at reasonable reaction rates.

Although the Tafel-slope may provide some qualitative information about the RDS, which DEMS can help elucidate, there is no doubt from a more practical aspect that the reaction kinetics i.e. activity and Tafel-slope are especially useful in the modelling and engineering of a fuel cell. The linearity of the Tafel-slope identifies the potential region in which the MOR is determined by the electrode kinetics, anything above or below can be considered blockage of the electrocatalyst surface [either via CO_{ads} (at lower potentials) or OH_{ads} species (at high potentials)]. For the Pt catalyst, the Tafel-slope region was observed to lie between 0.55 and 0.68 V_{RHE} whereas for the PtRu catalyst this region lie between 0.4 and 0.58 V_{RHE}, giving a narrow potential region where the activity of the two catalysts can be compared (i.e. between 0.55 and 0.58 V_{RHE}) in the following section.

5.3.6 Activity

The assessment and comparison of electrocatalytic activities is crucial toward the understanding of the fundamental relationship between activity and catalyst structure and toward the development or selection of suitable electrocatalyst materials. The activities of the mass and specific activities of the Pt and PtRu catalysts investigated in this study are summarised in Table 5.2.

Table 5.2 Summary of the Tafel-slopes and activities of the MOR on Pt 5 nm and PtRu electrocatalysts investigated in this study

	Method	Tafel slope/mV dec^{-1}	Potential/V_{RHE}					
			0.45	0.50	0.55	0.60	0.65	
Pt 5 nm	CV	82	-	-	-	-	19.4	mA mg_{metal}^{-1}
			-	-	-	-	0.030	mA cm_{real}^{-1}
	Chrono-A	96	-	-	1.74	5.77	19.2	mA mg_{metal}^{-1}
			-	-	0.003	0.009	0.030	mA cm_{real}^{-1}
PtRu	CV	134	19.4	47.5	107.5	-	-	mA mg_{metal}^{-1}
			0.021	0.052	0.116	-	-	mA mg_{real}^{-1}
	Chrono-A	129	13.3	30.1	72.6	-	-	mA mg_{metal}^{-1}
			0.010	0.024	0.053	-	-	mA cm_{real}^{-1}

Values were taken from within the potential window that exhibited Tafel-slope behaviour

These values were taken from the CV and chronoamperometric measurements at a number of potentials within the Tafel-slope potential region. The only overlap between the Tafel-slope regions of the MOR on the Pt and PtRu catalyst in these measurements is at 0.55 V_{RHE} using chronoamperometry, whereby the mass activity of the PtRu catalyst (72.6 mA mg^{-1}_{metal}) is almost two orders of magnitude greater than Pt (1.74 mA mg^{-1}_{metal}). If we then compare the potentials at ~19.4 mA mg^{-1}_{metal} of current we observe a ~200 mV improvement in the performance of the catalyst. Either way, it is clear that the PtRu catalyst is by far the more active and suitable MOR catalyst with measurements at electrode potentials ≥0.6 V_{RHE} bearing little relevance to the development of DMFC anodes. An issue arises, however, if we try to compare the MOR activities in this study with those of other catalyst materials based on the measurements of different research groups. It has in the past been noted that the comparison of MOR activities between different research groups were not unambiguous, citing differences in the 'history' of the electrodes, duration of the measurements, and/or different concentrations of the supporting electrolyte [35]. This can be considered an understatement of a situation that has not improved in the past 10 years, as is illustrated in Table 5.3 which gives details of the literature values and methodologies employed in the determination of MOR activities.

It is immediately clear that there are a number of variations in the electroanalytical techniques employed, which are performed in various different concentrations of electrolyte who probe the MOR at inconsistent WE potentials that are referenced to various different RE with a wide range of MOR activity values normalised in a number of different ways. The problem is not only identifying activities that are comparable to those in this study, but also identifying values from different studies that are comparable to each other because there is clearly no commonly accepted approach to determining and comparing the MOR activities of prospective catalyst materials.

This issue has been raised previously, although quite briefly in 1999, calling for the need of a standardised method for comparing prospective DMFC anode catalysts for the MOR [18]. In fuel cell related electrocatalyst research and development, these problems are unfortunately not isolated to the study of the MOR, but is also prevalent in, and previously highlighted for, the study of ORR activities [1]. This has since led to an effort to present and scrutinise the ORR methodology and offer suggested measurement guidelines [7], which were later improved upon in Ref. [16, 58] and demonstrated in Ref. [2], to facilitate the meaningful comparison between ORR activities of different electrocatalyst materials. In light of the experimental details presented in Table 5.3, it would appear that the same approach is urgently needed to facilitate meaningful comparison between MOR electrocatalysts, particularly where claims of novel materials with 'high' activity are made. The attention here was to highlight this issue rather than discuss ways in which it should be resolved (this is after all a DEMS study); however, some discussion of the issues in defining a methodology are given in Appendix A for future reference.

5.3.7 Potential Dependent Conversion

By investigating the MOR on HSAC supported -Pt and -PtRu catalyst using CV, sampled current chronopotentiometry and three-dimensional voltammetry combined with DEMS, a contrast in the potential dependence of the conversion of the MOR to CO_2 between the two catalysts became apparent, suggesting a key difference in the MOR mechanism and its potential dependence on each respective catalyst. Whereas on the Pt catalyst a close positive correlation was observed between the MOR electrode current and the conversion of methanol to CO_2, the conversion of methanol to CO_2 on the PtRu catalyst was found to be rather inversely correlated to the MOR electrode current. In order to discuss perhaps why we observed this inverse relationship on the PtRu electrocatalyst, it is necessary to first deliberate the possible explanations for the potential dependence of the MOR on the Pt catalyst in relation to the current understanding of the MOR mechanism.

In order for methanol to adsorb on the Pt surface and the MOR to proceed, the poisoning CO_{ads} species must be oxidise in order to free active sites for the dissociative adsorption of methanol (as presented in more detail in Sect. 5.1.1). The oxidation of the CO_{ads} species is in turn reliant on the bifunctional adsorption of and reaction with OH_{ads} species, whose adsorption on Pt has two opposing consequences on the MOR depending on OH_{ads} surface coverage. At low coverage, the OH_{ads} species can be considered as a promoter of the MOR by oxidising the primarily surface poisoning CO_{ads} species generated from the MOR, whereas at high surface coverage the OH_{ads} species themselves begin to block the electrode surface inhibiting the adsorption of methanol. The coverage of OH_{ads} species (which is strongly dependent on potential) is therefore crucial to the MOR that is understood to proceed via two generalised pathways: the direct pathway, whereby the MOR proceeds via a single dissociative adsorption step oxidising to CO_{ads} and then CO_2; and the in-direct pathway, where partial MOR products (such as formaldehyde and methylformate) desorb from the surface and may or may not re-adsorb and further oxidise to CO_2. In general, the calculated conversion of the

Table 5.3 Table summarising the MOR activities obtained from electrocatalytic studies on respective materials with relevant experimental details

Catalyst	Method	Temp/ C	MeOH Conc./ mol dm^{-3}	Electrolyte	Potential	Value	Unit	References
Ambient temperature								
Pt/C	Chrono–A	25	1.0	0.5, H_2SO_4	0.6 $V_{Ag/AgC}$	4.31	mA cm^{-2}	[36]
Pt/C	12 h					27.88		
Pt/C(E-Tek)						11.808		
Pt/C (Various % wt. Pt)	Chrono–A 30 min		0.1	0.1, $HClO_4$	0.7 V_{RHE}	~7–13	mA cm_{geo}^{-2}	[38][a]
Pt/C	Chrono–A 10 min		0.5	0.5, H_2SO_4	0.5 V_{RHE}	1.3	mA mg_{Pt}^{-1}	[33]
Pt_2Ru_3/C						1.4		
(Pt_{85}/Pt_{54})	CV 10 mV s^{-1}		0.1	0.5, H_2SO_4		~35	mA cm_{real}^{-2}	[8][a]
(Pt_{58}/Pt_{39})						~21		
Pt/C	CV 50 mV s^{-1}		1.0	1.0, H_2SO_4	1.0 $V_{Ag/AgCl}$	29.5	mA cm^{-2}	[50]
Pt-WO_3/C					0.9 V_{AgCl}	62.0		
Pd-TiO_2			1.0	1.0, H_2SO_4	0.7 V_{SCE}	1.1	mA	[51]
PtRu/Au	CV 20 mV s^{-1}		1.0	0.5, H_2SO_4	0.5 V_{NHE}	50	A $g_{PtRu^{-1}}$	[52][a]
PtRu/C						0.5		
$Pt_{53}Ru_{47}$/CNT	ChrA		2.0	1.0, H_2SO_4	0.5 V	4.60	molecule	[37]
$Pt_{69}Ru_{31}$/CNT						2.03	s^{-1}site^{-1}	
$Pt_{77}Ru$/CNT						3.05	$\times 10^{-3}$	
$Ni_{59}Nb_{40}Pt_1$	CV 50 mV s^{-1}		1.0	1.0, $HClO_4$	0.65 V_{NHE}	~0.05	mA $cm_{ECSA^{-2}}$	[53][a]
$Ni_{59}Nb_{40}Pt_{0.6}Sn_{0.4}$						~0.08		
						~0.1		

(Continued)

Table 5.3 (continued)

Catalyst	Method	Temp/ C	MeOH Conc./ mol dm^{-3}	Electrolyte	Potential	Value	Unit	References
$Ni_{59}Nb_{40}Pt_{0.6}Ru_{0.4}$								
Au–Pt (various)	CV 20 mV s^{-1}		1.0	0.5	0.6 V_{NHE}	~4–62	mA mg_{Pt}^{-1}	[54][a]
Pt NPs various	Chrono–A		0.5	0.5 H_2SO_4	0.5 V_{SCE}	~5–50	mA mg^{-1}	[39]
Raised temperature								
PtRu	Chrono–A	60	2.0	0.5 H_2SO_4	0.5 V_{RHE}	105	mA mg_{PtRu}^{-1}	[35]
TFE colloid			0.5			70		
Etek-Pt			0.5			15		
Etek-PtRu						50		
$Pt_{0.5}Ru_{0.5}$	Chrono–A		1.0	2.5 H_2SO_4		~80		[55][a]
$Pt_{0.5}Ru_{0.5}$	Chrono–A		2.0	1.5 H_2SO_4		~200		[56][a]
$Pt_{0.66}Ru_{0.34}$	Chrono–A		1.0	0.5 H_2SO_4		~210		[35][a]
$Pt_{53}Ru_{47}$/CNT	Chrono–A		2.0	1.0, H_2SO_4		26.3	molecule s^{-1} site^{-1} × 10^{-3}	[37]
$Pt_{69}Ru_{31}$/CNT						14.5		
$Pt_{77}Ru$/CNT						19.6		
MP-PtRu	Chrono–A		1.0	1 H_2SO_4	0.55 V_{RHE}	0.38	mA cm^{-2}	[40]
MP-Pt/Ru						0.42		
PtRu/C						0.05		
PtRu/TFC	CV, 10 mV s^{-1}		1.0	0.5 H_2SO_4	0.5 V_{RHE}	~230	mA mg_{metal}^{-1}	[57] [a]
PtRuP0.1/TFC						~375		
Pt (Various ternary)	Chrono–A, 60 min		1.0	1.0 H_2SO_4	0.6 V_{RHE}	~0.4–4	mA $mg_{cat.}^{-1}$	[41]

Note that the Ag/AgCl reference electrode potential is typically −0.25 V_{RHE}

References with [a] indicate data which was taken from figures because no specific values are given in either the text or a table

MOR on the Pt catalyst is far from complete, averaging at ~0.25 or less in these measurements, indicating that the majority of MOR is only partially oxidised under the experimental conditions suggesting that the MOR largely proceeds via the in-direct pathway. The partial MOR products, however, generally do not re-adsorb and react further to CO_2. Even so, the experimental results for the MOR reaction showed a strong correlation between the conversion to CO_2 and henceforth the oxidation of CO_{ads} and the coverage of OH_{ads} species. These results indicated that at rather high coverage of either the CO_{ads} or OH_{ads} species, the conversion was poorer than at the peak MOR potential.

If we consider that in the MOR we are not observing the charge transfer kinetics of the MOR because of the influence of adsorbed intermediates, but rather the distorted slope of the RDS and therefore the degree of poisoning of the Pt surface by either or both CO_{ads} or OH_{ads}, we can deduce that at the peak MOR current potential, the surface blockage is at a minimum and the most ensembles of Pt atoms required for either methanol or desorbed partial MOR products to adsorb (or re-adsorb) and oxidise are available. The minimised coverage of blocking species on the electrode could therefore conceivably promote both the direct and in-direct MOR pathways. For instance, a higher population of Pt ensembles required for the direct MOR pathway would be conceivably greater, whereas desorbed partial MOR products will also have more active sites available to re-adsorb and oxidise further (in competition with methanol adsorption). Each possibility may be equally valid and neither one can be discounted, although the majority of the MOR is still only partial oxidation, even at the peak current potential.

On PtRu the MOR mechanism was observed to be somewhat different. At ambient temperature, only the Pt component of the PtRu surface is understood to be active for the initial dissociative adsorption of methanol [6]. In order for the MOR to commence, Pt-CO_{ads} species must be oxidised, a process that is facilitated in the bifunctional reaction with adsorbed Ru-OH_{ads} species at low potentials. Again, the methanol may dissociatively adsorb reacting via either the direct or in-direct pathway to CO_2. The measurements in this study, however, indicate that the conversion at low potentials is rather high despite that the MOR rate is fairly low which would suggest that the coverage of Pt-CO_{ads} species, at least on the active site for the dissociative adsorption of methanol is considerable. This is in contrast to the observations on the pure Pt catalyst and it would seem unlikely that either of the MOR pathways would be favoured at low potentials, until we consider the possible secondary role of the Ru component of the PtRu surface. Although Ru is not active for the initial adsorption of methanol at room temperature [3, 4], pure Ru electrodes have been observed to possess some activity for the oxidation of both formaldehyde [29] and formic acid [4] partial MOR products. It is therefore proposed that the role of Ru in the MOR under these experimental conditions is more than purely providing the OH_{ads} species necessary to oxidise the CO_{ads} on Pt at low potentials, but is to also oxidise partial MOR products that desorb from the Pt surface in the in-direct MOR pathway. The decrease in the conversion with increasing MOR rate and potential on the PtRu catalyst may then be rationalised in two ways. Firstly, the increasing MOR rate on the Pt surface will generate a larger number of desorbed partial MOR products which will eventually exceed the oxidising capacity of the Ru surface. Secondly, the ability of the Ru to oxidise these intermediates will also change (similarly to Pt), initially increasing as OH_{ads} are adsorbed until at a certain point where it peaks and diminishes as the coverage of OH_{ads} species increases. Although the ability of the Ru component to oxidise partial MOR products has in general been neglected in the discussions of the MOR mechanism in the scientific literature, there is much evidence to support it. For example, formic acid oxidation on pure Ru electrode commences at ~0.35 V_{RHE} peaking at 0.5 V_{RHE} before decreasing with increasing electrode potential [4]. This is closely related to the adsorption of OH_{ads}, which has be probed by studying the blocking effect of OH_{ads} on the HOR on pure Ru electrodes, whereby the HOR reaction decreases between 0.2 and 0.5 V_{RHE} as OH_{ads} species form on the Ru surface [59] analogous to the effect of OH_{ads} on the HOR on pure Pt electrodes at 0.9 V_{RHE} [16]. If we furthermore consider that Ru is not active for the initial dissociative adsorption of methanol <60 °C, the re-adsorption of desorbed partial MOR products do not compete with methanol for the adsorption on active sites on Ru, in contrast to competitive re-adsorption on Pt (assuming the active sites for both methanol and partial MOR products on Pt are the same). The overall conversion is therefore determined by the relative rates of the predominantly partial MOR on the Pt surface to produce desorbed intermediates, and their subsequent re-adsorption and oxidation on either Pt (in competition with methanol) or Ru.

These finding will have less significance to electrochemical half-cell measurements or DMFCs which operate at temperatures >60 °C where methanol is acknowledged to adsorb on a pure Ru electrode and therefore generally participate more in the MOR. The measurements do, however, highlight that rather high conversion of the MOR can be observed on the PtRu catalyst under these experimental conditions, albeit at low MOR overpotentials and rates. The findings may have more relevance to portable DMFC devices that operate close to room temperature; however, the conversion of the MOR to CO_2 may not have major significance because partial MOR products will still eventually be oxidised to CO_2, unless the altered composition of the DMFC fuel is a major practical problem (for example, methylformate may cross over the membrane to the cathode more easily than methanol). The findings at least serve to clarify the difference in the MOR mechanisms for Pt and PtRu that occur during the measurement in the fundamental studies in electrochemical half-cells, indicating that the MOR mechanism is dependent on the electrode potential and highlighting that it should not be overlooked that the role of Ru is more than just the promotion of the MOR via the adsorption of OH_{ads} species.

5.4 Conclusions

This study set out to not only demonstrate the research capabilities of the DEMS instrument, but to also re-visit the MOR on HSAC supported Pt and PtRu catalysts. Although the MOR has been extensively studied for many years using a variety of techniques, DEMS is still able to provide new insights into the MOR, demonstrating the contrast between MOR mechanisms on Pt and PtRu catalysts. Furthermore, contrary to previous DEMS studies the CO-Stripping charge of the PtRu catalyst was found not to be overestimated if the background CV in Argon saturated electrolyte is subtracted prior to the integration of the CO-Stripping current.

An extremely close relationship was observed between the LSV of a saturated CO_{ads} layer, and the MOR electrode current and corresponding I_{CO2} on the Pt catalyst. The peak conversion of the MOR to CO_2 of $\sim$0.35 corresponded precisely with the CO-Stripping and MOR electrode current peak potential at 0.815 V_{RHE}. Furthermore, the Tafel-slopes for the MOR electrode current and corresponding I_{CO2} were seen to be almost identical at $\sim$80 mV Dec^{-1} during CV and $\sim$96 mV Dec^{-1} using sampled current chronoamperometry. This observation is in accordance to the widely accepted understanding that the oxidation of the poisoning CO_{ads} species is the RDS for the MOR on pure Pt catalysts. The close relationship between the MOR rate and CO_{ads} oxidation on the Pt catalysts, however, was not mirrored by the observations of the MOR on the PtRu catalyst. On the PtRu catalyst, the highest conversion of the MOR to CO_2 was observed at rather low reaction rates, decreasing from >0.7 at 0.4 V_{RHE} to $\sim$0.2 at 0.6 V_{RHE}. Furthermore, the Tafel slope of $\sim$130 mV Dec^{-1} for the electrode current of the MOR was not reflected in the I_{CO2} whose slope was $\sim$200 mV Dec^{-1} with significantly less linearity. This would support the interpretation that the oxidation of CO_{ads} is not the RDS of the MOR on PtRu catalysts but rather the initial dissociative adsorption of methanol on Pt, although the poisoning CO_{ads} species can still be attributed to the significant electrode overpotential required for the MOR to proceed. The trends in the MOR mechanism were furthermore confirmed over the entire current, time and potential relationship of the MOR measurement using 3-dimensional voltammetry, which was for the first time used to study the MOR in combination with DEMS. The resulting current contour plots provided a visual aid of the MOR system by encapsulating the current, potential and time relationship contained in multiple chronoamperometric within a single interpretable plot. These plots revealed changes in the MOR mechanism on the PtRu catalyst over time, during chronoamperometric measurements at $\geq$0.55 V_{RHE} and was attributed to potential induced changes in the surface morphology of the PtRu catalyst.

The mass activity of the MOR on the PtRu catalyst (72.7 mA mg_{metal}^{-1}) meanwhile was observed to be almost two orders of magnitude greater than the Pt catalysts (1.74 mA mg_{metal}^{-1}) at the comparable potential of 0.55 V_{RHE} using chronoamperometry. The comparison of these values to other studies was, however, found to be extremely ambiguous because there is no commonly accepted and applied methodology to assess the activity of prospective MOR catalysts materials. These inconsistencies cannot be intuitive to MOR electrocatalyst development, particularly where claims of 'high activity' on novel materials are made, and guidelines analogous to those suggested for the ORR are proposed to be beneficial, although for the MOR there are certainly greater difficulties in defining guidelines in comparison to the ORR methodology using RDE.

Perhaps the most significant insight into the MOR of this study was the contrast in the potential dependent conversion of the MOR to CO_2on Pt and PtRu catalysts. On Pt the conversion of the MOR to CO_2 was directly related to its ability to oxidise CO_{ads} species. The contrasting trend on the PtRu catalyst, where relatively high conversion was observed at low MOR rates (<0.6 V_{RHE}) and ambient temperature ($\sim$298 K) was attributed to the ability of Ru to oxidise desorbed partial MOR reaction products such as formaldehyde and formic acid that arise from the partial MOR on the Pt component of the PtRu surface. The decrease in conversion with increasing potential on PtRu was rationalised by the difference in the relative rates of the MOR on the Pt, and the re-adsorption and oxidation of partial MOR products on either Pt (in competition with methanol) or Ru.

Overall, these findings serve to provide greater insight into the electrochemical half-cell study of the MOR, highlighting that the ability of Ru to oxidise partial MOR products may be significant at low overpotentials allowing rather high MOR conversion to CO_2 at ambient temperature.

References

1. Gasteiger HA et al (2005) Activity benchmarks and requirements for Pt, Pt-alloy, and non-Pt oxygen reduction catalysts for PEMFCs. Appl Catal B 56(1–2):9–35
2. Wiberg GKH, Mayrhofer KJJ, Arenz M (2010) Investigation of the oxygen reduction activity on silver—a rotating disc electrode study. Fuel Cells 10(4):575–581
3. Gasteiger HA et al (1993) Methanol electrooxidation on well-characterized platinum-ruthenium bulk alloys. J Phys Chem 97(46):12020–12029
4. Markovic NM et al (1995) Electrooxidation mechanisms of methanol and formic-acid on pt-ru alloy surfaces. Electrochim Acta 40(1):p 91–98
5. Gasteiger HA et al (1994) Temperature-dependent methanol electrooxidation on well-characterized pt-ru alloys. J Electrochem Soc 141(7):1795–1803
6. Gasteiger HA et al (1994) Electrooxidation of small organic-molecules on well-characterized pt-ru alloys. Electrochim Acta 39(11–12):1825–1832
7. Mayrhofer KJJ et al (2008) Measurement of oxygen reduction activities via the rotating disc electrode method: from pt model surfaces to carbon-supported high surface area catalysts. Electrochim Acta 53(7):3181–3188
8. Jusys Z, Kaiser J, Behm R.J et al (2002) Composition and activity of high surface area PtRu catalysts towards adsorbed CO and methanol electrooxidation—a DEMS study. Electrochim Acta, 47(22–23):3693–3706
9. Jusys Z, Behm RJ (2001) Methanol oxidation on a carbon-supported pt fuel cell catalysta kinetic and mechanistic study by differential electrochemical mass spectrometry. J Phys Chem B 105(44):10874–10883
10. Jambunathan K, Jayaraman S, Hillier AC (2004) A multielectrode electrochemical and scanning differential electrochemical mass spectrometry study of methanol oxidation on electrodeposited ptxruy. Langmuir 20(5):1856–1863
11. Jusys Z, Kaiser J, Behm RJ (2003) Methanol electrooxidation over pt/c fuel cell catalysts: dependence of product yields on catalyst loading. Langmuir 19(17):6759–6769
12. Willsau J, Wolter O, Heitbaum J (1985) On the nature of the adsorbate during methanol oxidation at platinum—a dems study. J Electroanal Chem 185(1):163–170
13. Jusys Z et al (2002) Activity of PtRuMeOx (Me = W, Mo or V) catalysts towards methanol oxidation and their characterization. J Power Sources 105(2):297–304

14. Wang H, Baltruschat H (2007) DEMS study on methanol oxidation at poly—and monocrystalline platinum electrodes: the effect of anion, temperature, surface structure, ru adatom, and potential. J Phys Chem C 111(19):7038–7048
15. Wang H, Loffler T, Baltruschat H (2001) Formation of intermediates during methanol oxidation: a quantitative dems study. J Appl Electrochem 31(7):759–765
16. Wiberg GKH (2010) The devlopment of a state-of-the-art experimental setup demonstrated by the investigation of fuel cell reactions in alkaline electrolyte, in Lehrstühl für Physicalische Chemie. Technische Universität München: München
17. Dillon R et al (2004) International activities in DMFC R&D: status of technologies and potential applications. J Power Sources 127(1–2):112–126
18. Wasmus S, Küver A (1999) Methanol oxidation and direct methanol fuel cells: a selective review. J Electroanal Chem 461(1–2):14–31
19. Wang K et al (1996) On the reaction pathway for methanol and carbon monoxide electrooxidation on Pt-Sn alloy versus Pt-Ru alloy surfaces. Electrochim Acta 41(16):2587–2593
20. Gasteiger HA et al (1994) Carbon monoxide electrooxidation on well-characterized platinum-ruthenium alloys. J Phys Chem 98(2):617–625
21. Serov A, Kwak C (2009) Review of non-platinum anode catalysts for DMFC and PEMFC application. Appl Catal B 90(3–4):313–320
22. Lai S et al (2007) Mechanisms of carbon monoxide and methanol oxidation at single-crystal electrodes. Top Catal 46(3):320–333
23. Gojkovic SL (2003) Electrochemical oxidation of methanol on Pt3Co bulk alloy. J Serb Chem Soc 68(11):11
24. Wang H et al (2001) Methanol oxidation on Pt, PtRu, and colloidal Pt electrocatalysts: a DEMS study of product formation. J Electroanal Chem 509(2):163–169
25. Christensen PA et al (1994) An in situ FTIR study of the electrochemical oxidation of methanol at small platinum particles. J Electroanal Chem 370(1–2):251–258
26. Korzeniewski C, Childers CL (1998) Formaldehyde yields from methanol electrochemical oxidation on platinum. J Phys Chem B 1998 102(3):489–492
27. Islam M, Basnayake R, Korzeniewski C (2007) A study of formaldehyde formation during methanol oxidation over PtRu bulk alloys and nanometer scale catalyst. J Electroanal Chem 599(1):31–40
28. Batista EA, Iwasita T (2006) Adsorbed intermediates of formaldehyde oxidation and their role in the reaction mechanism. Langmuir 22(18):7912–7916
29. de Lima RB et al (2007) Catalysis of formaldehyde oxidation by electrodeposits of PtRu. J Electroanal Chem 603(1):142–148
30. Kucernak AR et al (1998) Anodic oxidation of methyl formate and its relationship to the reactions of methanol and formic acid. Electrochim. Acta 43(12–13):1705–1714
31. Wang HS et al (2001) Methanol oxidation on Pt, PtRu, and colloidal Pt electrocatalysts: a DEMS study of product formation. J Electroanal Chem 509(2):163–169
32. Lin WF et al (2000) Electrochemical versus gas-phase oxidation of ru single-crystal surfaces. J Phys Chem B 104(25):6040–6048
33. Tripkovic AV et al (2002) Methanol electrooxidation on supported Pt and PtRu catalysts in acid and alkaline solutions. Electrochim Acta 47(22–23):3707–3714
34. Tripkovic AV et al(2006) Methanol oxidation at platinum electrodes in acid solution: comparison between model and real catalysts. vol 71 SERBIE: serbian chemical society, Belgrade p 11
35. Schmidt TJ, Gasteiger HA, Behm RJ (1999) Methanol electrooxidation on a colloidal PtRu-alloy fuel-cell catalyst. Electrochem Commun 1(1):1–4
36. Teng Z-H, et al (2007) High activity Pt/C catalyst for methanol and adsorbed CO electro-oxidation. J Power Sources 164(1):105–110
37. Li L, Xing Y (2009) Methanol electro-oxidation on pt-ru alloy nanoparticles supported on carbon nanotubes. Energies 2(3):789–804

38. Bergamaski K et al (2006) Nanoparticle size effects on methanol electrochemical oxidation on carbon supported platinum catalysts. J Phys Chem B 110(39):19271–19279
39. Qiu L et al (2011) Fabrication of ionic liquid-functionalized polypyrrole nanotubes decorated with platinum nanoparticles and their electrocatalytic oxidation of methanol. Chem Commun 47(10):2934–2936
40. García G et al (2011) Methanol electrooxidation at mesoporous Pt and Pt-Ru electrodes: a comparative study with carbon supported materials. J Power Sources 196(6):2979–2986
41. Jeon MK, McGinn PJ (2009) Improvement of methanol electro-oxidation activity of PtRu/C and PtNiCr/C catalysts by anodic treatment. J Power Sources 188(2):427–432
42. Gojkovic SL (2004) Mass transfer effect in electrochemical oxidation of methanol at platinum electrocatalysts. J Electroanal Chem 573(2):271–276
43. Papadopoulos N et al (1991) Three-dimensional electrochemistry: utilization of i-E-t curves for elucidation of electrochemical reactions. J Electroanal Chem 308(1–2):83–96
44. Yang R et al (2005) Monodispersed hard carbon spherules as a catalyst support for the electrooxidation of methanol. Carbon 43(1):11–16
45. Maillard F, Savinova ER, and Stimming U (2007) CO monolayer oxidation on Pt nanoparticles: further insights into the particle size effects. J Electroanal Chem 599(2):221–232
46. Sarma LS et al (2007) Investigations of direct methanol fuel cell (DMFC) fading mechanisms. J Power Sources 167(2):358–365
47. Gojkovic SL, Vidakovic TR, Durovic DR (2003) Kinetic study of methanol oxidation on carbon-supported PtRu electrocatalyst. Electrochim Acta 48(24):3607–3614
48. Meli G et al (1993) Direct electrooxidation of methanol on highly dispersed platinum-based catalyst electrodes—temperature effect. J Appl Electrochem 23(3):197–202
49. Noel M, Vasu KI (1990) Cyclic voltammetry and the frontiers of electrochemistry. South asia books
50. Maiyalagan T, Viswanathan B (2008) Catalytic activity of platinum/tungsten oxide nanorod electrodes towards electro-oxidation of methanol. J Power Sources 175(2):789–793
51. Wang M, Guo D-j, Li H-l (2005) High activity of novel Pd/TiO2 nanotube catalysts for methanol electro-oxidation. J Solid State Chem 178(6):1996–2000
52. Lee K-S et al (2009) PtRu overlayers on Au nanoparticles for methanol electro-oxidation. Catal Today 146(1–2):20–24
53. Barranco J, Pierna AR (2007) Bifunctional amorphous alloys more tolerant to carbon monoxide. J Power Sources 169(1):71–76
54. Park I-S et al (2007) Electrocatalytic activity of carbon-supported Pt-Au nanoparticles for methanol electro-oxidation. Electrochim Acta 52(18):5599–5605
55. Hamnett A et al (1990) Long-term poisoning of methanol anodes. Ber Bunsen Ges Phys Chem 94(9):1014–1020
56. Watanabe M, Uchida M, Motoo S (1987) Preparation of highly dispersed Pt + Ru alloy clusters and the activity for the electrooxidation of methanol. J Electroanal Chem Interfacial Electrochem 229(1–2):395–406
57. Lin M-L, Lo M-Y, Mou C-Y (2011) PtRuP nanoparticles supported on mesoporous carbon thin film as highly active anode materials for direct methanol fuel cell. Catal Today 160(1):109–115
58. Wiberg GK, Mayrhofer KJ, Arenz M (2009) Investigation of the oxygen reduction activity of non-platinum catalysts: a RDE methodology. ECS Meet Abs 902(6):349
59. Gasteiger HA, Markovic NM, Ross PN (1995) H2 and CO electrooxidation on well-characterized Pt, Ru, and Pt-Ru. 1. Rotating disk electrode studies of the pure gases including temperature effects. J Phys Chem 99(20):8290–8301

Chapter 6
The Electrochemical Oxidation of HSAC Catalyst Supports

A fundamental understanding of the electrochemical oxidation behaviour of carbon blacks used to support finely dispersed Pt particles is crucial to the design of mitigating strategies that prevent the deterioration of PEMFC performance, particularly in automotive applications. In this study, DEMS is utilised to investigate and compare the electrochemical oxidation tendencies of a pristine EC300 carbon black and five graphitised carbon blacks (heat-treated between 2,100 and 3,200 °C). The study is then briefly extended to EC300 HSAC supported Pt nanoparticle catalysts. By monitoring the CO_2 and O_2 produced in situ, it is possible to elucidate the contributions of the different electrochemical reactions to the overall faradaic electrode current observed during high potential excursions $>1.35\ V_{RHE}$. The initial partial and complete electrochemical oxidation of EC300 was found to be considerably greater than the graphitised samples. However, during repeated potential excursions the electrochemical oxidation tendencies were observed to be dynamic, exhibiting three trends that are not only influenced by the amount of electrochemical oxidation but also the lower electrode potential limit. Finally, the behaviour of the EC300 HSAC supported Pt catalysts was found to be largely the same as the pure EC300 carbon black, however, an additional peak is observed after electrochemical oxidation attributed to the oxidation of CO species on Pt.

6.1 Introduction

The key aspects of PEMFC research and development are focused towards resolving issues concerning the cost, efficiency and durability of the CEEC device. Although every component of the PEMFC contributes to at least one of these aspects (to variable extents) the most fundamentally crucial is the electrocatalyst material of the MEA. In terms of durability, several processes such as PEM

S. J. Ashton, *Design, Construction and Research Application of a Differential Electrochemical Mass Spectrometer (DEMS)*, Springer Theses,
DOI: 10.1007/978-3-642-30550-4_6, © Springer-Verlag Berlin Heidelberg 2012

thinning, gas diffusion layer degradation, electrocatalyst (Pt) particle sintering and HSAC support corrosion have been found to contribute to the PEMFC performance deterioration [1], of which the latter two factors are directly related to the characteristics of the electrocatalyst material. In particularly demanding PEMFC applications such as in an automobile, the degradation in the performance is considerable [2] and consequently, a thorough understanding of the failure mode(s) for each FC component in the development mitigating strategies to improve PEMFC endurance are urgently required [1]. Of particular concern to the PEMFC lifetime is the number of start-stop events [2–7] during which the potential of the cathode may reach potentials up to 1.75 V_{RHE} [6] evolving considerable quantities CO_2 [8] corroding the electrocatalyst HSAC support and rapidly deteriorating the PEMFC performance.

There are essentially two types of approach to the study of PEMFC degradation processes: (1) lifetime testing (LT) and (2) accelerated stress testing (AST) [1, 9]. LT involves operating the PEMFC as it is intended to be used, which although certainly provides the most reliable indication of durability, is both time consuming and expensive with slow accumulation of data. The interpretation of LT data is also complicated by the wide range of parameters that can influence the long term durability PEMFCs [10]. Consequently, AST is a considerably more popular approach to assessing PEMFC durability, simply because data can be accumulated relatively quickly, under controlled conditions leading to more rapid developments. These ASTs can (similarly to the approach in assessing PEMFC electrocatalyst activities as described in Chap. 1) take two forms, and involve either in situ or ex-situ methodologies.

In situ AST methods typically involve fabricating an MEA and subjecting it to specific treatments under controlled conditions that push the boundaries of the intended operating region of the PEMFC in order to stress and accelerate the degradation of the MEA components. In such experiments the performance of the MEA is often monitored as a function of the AST parameter(s), such as current load, current load cycles or operating temperature [2, 7, 11–13]. The effect of the AST treatment on the MEA components may then be assessed by analysing the condition of the MEA after a degradation treatment [11, 13–15], identifying process such as cathode thinning, Pt particle loss and Pt precipitation [14, 15]. There are also experiments which monitor corrosion processes online, for example, by measuring and correlating the CO_2 concentration in cathode PEMFC exhaust gas using mass spectrometry [16, 17] or sampled gas chromatography [7, 12] with the deteriorating in the PEMFC performance. Although undoubtedly essential, in situ methods are still relatively expensive and require the control of a large number of experimental variables (temperature, pressure, humidity etc.) and fabrication of the MEA. The screening of electrocatalysts or HSAC support materials, as well as the precise elucidation of degradation processes to specific PEMFC components (whose deterioration may not be uniform but localised [13]) is by no means trivial. A complementary approach using ex-situ AST techniques are therefore useful in order to enhance our fundamental understanding of the degradation processes of specific components, and to screen prospective HSAC support materials to facilitate the development of mitigating strategies that enhance the PEMFC lifetime.

Ex-situ AST methods involve assessing the individual component of an MEA, such as the study of the electrocatalyst material in a three-electrode electrochemical half-cells, which have been used to measure the electrochemical oxidation of model carbons [18, 19], carbon blacks [20–22] and HSAC supported Pt catalysts [20, 23, 24]. These studies typically involve applying a certain electrochemical treatment, using either CV [23–25], chronoamperometry [20] or potential pulse [20, 26] perturbation to a potential of at least 1.2 V_{RHE}, with the intention of electrochemically oxidising the carbon or electrocatalyst material. The anodic electrode current observed is then often referred to as corrosion current [21] or it is assumed to correspond with the loss of carbon material [20]. The electrochemical AST treatment is then compared to a measurable property such as the apparent electrode double-layer capacitance [20] or the nature of the surface oxygen containing functional groups [26]. In electrochemical half-cell studies of HSAC supported Pt catalyst corrosion, however, the analysis of the electrode current is further complicated by the electrochemical reactions of the supported Pt particles, in particular the oxidation of the Pt surface >0.74 V_{RHE} and the oxygen evolution reaction (OER) >1.5 V_{RHE}. Using standard electrochemical half-cell methods therefore, it is impossible to decipher the carbon oxidation current based on the electrode current alone. Consequently, a measurable property of the electrocatalyst such as the ECSA of the supported Pt particles is often related to a parameter of the corrosion treatment, such as number of CVs [23–25], potential pulses [18] or potential hold time [20, 27], analogous to in situ AST approaches. These methodologies have also been enhanced by utilising analytical techniques such as transmission electron microscopy (TEM) [23, 24], secondary electron microscopy (SEM) [19], atomic force microscopy (AFM) [25, 27], X-ray photoelectron microscopy (XPS) [18, 25, 27] and EQCM [22]. For example, non-destructive TEM has been used to visually analyse the changes that occur to a specific electrocatalyst particle following a certain electrochemical treatment, identifying particle loss as the major cause of ECSA decrease during potential cycles up to 1.4 V_{RHE} at 25 °C. Alternatively, corrosion half-cell studies may also focus on the dissolution pure Pt electrodes [28].

Standard electrochemical half-cell techniques are nevertheless limited to the clouded interpretation of a single electrode current, even though a number of different electrochemical reaction processes are known to occur in the high potential regions >1.2 V_{RHE}, particularly in the study of HSAC supported Pt catalysts. The DEMS technique is therefore potentially very useful in the fundamental study of electrochemical oxidation processes by quantitatively measuring CO_2 and O_2, and has already been demonstrated in a handful of qualitative corrosion studies [29–31].

6.1.1 Motivation

The intention of this study was to investigate the electrochemical oxidation tendencies of a number of HSAC supports (a.k.a. carbon blacks), and to develop a methodology in which DEMS can be used to study and compare their carbon oxidation reaction (COR) processes effectively. In order to approach this problem, a comparison using CV is made between the electrochemical oxidation tendencies of a pristine EC300 HSAC and five graphitised samples (of the same raw carbon black material) prepared at various different temperatures between 2,100 and 3,200° C. Rather than focusing this study on electrode potentials within a PEMFC relevant region to observe effects of a quite subtle nature and attempt to interpret them claiming direct PEMFCs relevance, this study will instead probe relatively high electrode potentials $\leq$2.05 V_{RHE} with the intention of revealing clear trends in the electrochemical oxidation using DEMS. The electrochemical treatment will demonstrate the contrasting electrochemical oxidation tendencies of the various HSAC support samples to repeated high potential excursions, along with the possibly significant influence of the lower electrode potential limit. The study is then briefly extended to the electrochemical oxidation of EC300 HSAC supported Pt nanoparticle catalysts. Using DEMS, we are able to elucidate the contributions of the partial and complete COR processes, and the OER to the observed faradaic electrode current, relating the HSAC weight loss to the Pt ECSA loss. The methodology and experimental findings presented in this study should offer a valuable foundation for future DEMS amongst other electrochemical half-cell studies on the electrochemical oxidation processes of HSAC supports.

6.1.2 Background

The HSACs that are commonly used to support finely dispersed Pt particles are grouped in a classification of material termed 'carbon blacks' which are manufactured via the incomplete combustion (or thermal decomposition) of hydrocarbons under controlled conditions. Carbon blacks are rather high purity carbonaceous materials (typically $\geq$97 % wt. carbon compared to the $\sim$60 % carbon content of ash) that possess distinct, although wide ranging physiochemical properties that are useful in a number of industrial applications. In electrochemistry, carbon blacks possess the useful combination of acceptable electron conductivity, along with chemical and electrochemical compatibility with the electrochemical environment. The amorphous carbon black materials are considered to be formed from sections of graphene layers with little order parallel to the layer planes [32]. A carbon black particle possesses an aciniform structure (like a 'bunch of grapes'), which retains greater order of the graphene planes in the outer layers of the microstructure compared to the more disordered core. The high surface area and compatibility of carbon blacks with the electrochemical environment make them almost ideal supports for finely dispersed Pt electrocatalyst particles; however, they are inherently prone to being electrochemically oxidised and consumed in the PEMFC environment, deteriorating the electrocatalyst performance.

The complete oxidation of HSAC supports to CO_2 may be described by the following overall reaction mechanism:

$$C + 2H_2O \rightarrow CO_2 + 4H^+ + 4e^- \quad E^0 = 0.207\ V_{SHE}$$

The low standard electrode potential of 0.207 V_{SHE} would suggest that carbon would be rather unstable in the fuel cell environment, however, because of the high irreversibility of the reaction, the complete oxidation of carbon remains relatively slow within typical PEMFC operating potentials i.e. $\leq 1.2\ V_{RHE}$ [33]. Although the study of HSAC support corrosion in the PEMFC environment has only relatively recently received greater attention [34], the corrosion of HSAC supports in the phosphoric acid fuel cell (and half-cells aqueous electrolyte) is comparably well-studied owing to the more corrosive environment (due to higher operating temperatures). These earlier studies observed that the electrochemical oxidation of carbon involves both the complete oxidation to CO_2 or partial oxidation of the carbon to either oxygen containing surface functional groups or CO [26, 35]. Although a number of mechanisms have been proposed to explain the COR processes, none are considered complete and can explain all the experimental observations [35], which often produce apparently contradictory results, as discussed in Ref. [21]. The carbon corrosion mechanism(s) therefore remain rather speculative, although the inconsistent findings could be attributed to differences in experimental conditions, use of inconsistent terminology and both of the complexity of the COR reaction processes and the difficulties in characterising the inhomogeneous carbon black materials as discussed in Ref. [21]. The COR processes that occur on carbon black supports are determined by a number of concurrent phenomena that are influenced by the bulk and surface characteristics of the carbon material [21]. Crucial bulk properties include the d-spacing (a measure of the graphitic character) and BET area that are each determined by the microstructure of the carbon black, whereas the surface characteristics are defined by the nature and concentration of surface functional groups. Although there are a number of classifications of functional groups present on the surface of the carbon blacks, those containing oxygen are undoubtedly the most significant. These oxygen containing functional groups not only determine many of the physiochemical properties of the carbon blacks, but also strongly influence the COR processes and may be classified into three groups: acidic, basic and neutral, each possessing different reactivities. Acidic oxygen functional groups include carboxylic or weaker acids such as alcohols and hydroquinones, whereas neutral functional groups include quinones and lactones. Basic surface oxygen functional groups meanwhile are thought to resemble heterocyclic pyran rings [36].

A crucial advantage of carbon blacks is that by controlling the synthesis and/or subsequent treatment, it is to some extent possible to tune desirable properties of

Table 6.1 Details of the heat-treatment temperature, BET surface area and corresponding gravimetric and specific double layer capacitances for each of the carbon black samples investigated in this study.

Sample	Treatment temp./ °C	BET area/ $m^2\ g^{-1}$	Gravimetric capacitance/ F g^{-1}	Capacitance per BET area/ $\mu F\ cm_{BET}^{-2}$
EC300	–	720	61.2 ± 4.0	8.5
(4)	2,100	247	14.1 ± 0.8	5.7
(3)	2,450	230	10.6 ± 1.5	9.4
(2)	2,800	218	7.5 ± 1.0	3.4
(5)	3,000	195	13.9 ± 1.4	7.1
(1)	3,200	135	8.5 ± 0.9	6.2

the material toward its intended application. For example, heat-treatment in an inert atmosphere between ca. 1,000 and 2,700 °C is often used to graphitise raw carbon blacks, which essentially induces the stacking and order of the graphene layers decreasing the amorphous characteristics of the carbon microstructure. This treatment is an effective method of tuning the characteristics of the carbon black because between the two extremes forms of carbon: graphite and amorphous, there are a wide range of properties. Graphitisation, however, not only affects the structure of the carbon blacks but also changes the nature and concentration of surface functional groups. For instance, even mild heat-treatment <450° C is able to reduce the surface concentration of carboxylic acid, carbonyl and lactone surface functional groups, which are directly related to the volatile content of the carbon black [37]. The influence of heat treatment, however, will of course be dependent on both the bulk and surface characteristics of the raw carbon black material, and the heat-treatment temperature, atmosphere and duration.

6.2 Experimental

Each carbon black stock suspension was prepared by ultrasonically dispersing the powdered HSAC (EC300 and 5 heat-treated) in Millipore Milli-Q water to give solution concentration of 2.5 $\mu g_c\ \mu l^{-1}$ (EC300) and 4.3 $\mu g_c\ \mu l^{-1}$ (Samples 1–5). Details of the heat-treated (graphitised) samples, heat treatment temperatures and BET areas are presented later in Sect. 6.4.1 Table 6.1. Each measurement sample was then prepared by pipetting 20 μL of an aqueous catalyst suspension onto a mirror polished 5 mm Ø polycrystalline Au WE substrate and dried in air to provide sample loadings of 50 $\mu g_c\ cm^{-2}$ (EC300) and 85 $\mu g_c\ cm^{-2}$ (Samples 1–5). The EC300 HSAC supported Pt nanoparticle catalyst samples were meanwhile prepared in an analogous manner to provide carbon loadings of 50 $\mu g_c\ cm^{-2}$. The particle sizes and % wt. Pt of the HSAC supported nanoparticles catalysts are as follows: 1 nm–19.4 %, 3 nm–46.0 % and 5 nm–50.6 %.

The DEMS electrochemical flow cell setup used in this study is identical to that described previously in Sect. 5.2. All experiments were performed at room temperature (295 K) in aqueous 0.5 mol dm^{-3} H_2SO_4 (Normaton; Merck, Germany) electrolyte prepared in Millipore Milli-Q water (<18.3 MOhm cm^{-1}, TOC < 5 ppb). The DEMS cell collection efficiency determined in each experiment using CO-Stripping on a polycrystalline Pt electrode, and the QMS was calibrated for CO_2 and O_2 at $^{m}/_{z} = 44$ and $^{m}/_{z} = 32$, respectively. The evolution of $I_{m/z\,=\,44}$ and $I_{m/z\,=\,32}$ was followed online by the QMS, and the faradaic current corresponding to the production of CO_2 and O_2 was calculated using the respective calibration constants K^0 and measured cell collection efficiency N.

The I_{CO2} currents are calculated from the $I_{m/z\,=\,44}$ using the following equation:

$$I_{CO_2} = \frac{I_{m/z=44}4F}{K^0_{CO_2}N_{CO_2}}$$

The values for K^0_{CO2}, was 2.95 C mol^{-1}, and N was typically ~0.30 at a flow rate of 5 μl s^{-1}. A total of four electrons are assumed to be produced for each CO_2 evolved. The detection limit of the DEMS instrument for CO_2 is 2.754 ug_c h^{-1} and corresponds to ~25 μA cm^{-2} of faradaic current (assuming 4 electrons per CO_2 molecule).

The equivalent faradaic current arising from the oxygen evolution reaction was calculated using the following equation:

$$I_{O_2} = \frac{I_{m/z=32}4F}{K^0_{O_2}N_{CO_2}}$$

The values for K^0_{O2}, was 1.61 C mol^{-1} and we assume that the N_{O2} is identical to that of N_{CO2} which are found to be experimentally the same, within the error of the DEMS calibration measurement.

The remaining anodic electrode charge that does not arise from either the complete COR of the HSAC support or the OER on the polycrystalline Au background was assumed to arise from the partial COR, according to the following equation:

$$Q_{Partial\ COR} = Q_F - Q_{CO_2} - Q_{O_2}$$

All electrode potentials meanwhile are given versus the RHE determined using the HOR/HER in hydrogen saturated electrolyte, measured using a polycrystalline Pt electrode prior to experimentation.

6.2.1 Electrochemical Oxidation Procedure

In accordance with the objective of this study, the electrochemical measurement procedure was designed with the intention of obtaining an overview of the electrochemical oxidation tendencies of the carbon black samples. Prior to the electrochemical oxidation treatment, the samples were first characterised using CV between 0.05 and 1.0 V_{RHE} at various scan rates between 0.02 and 1.0 V s^{-1}. The

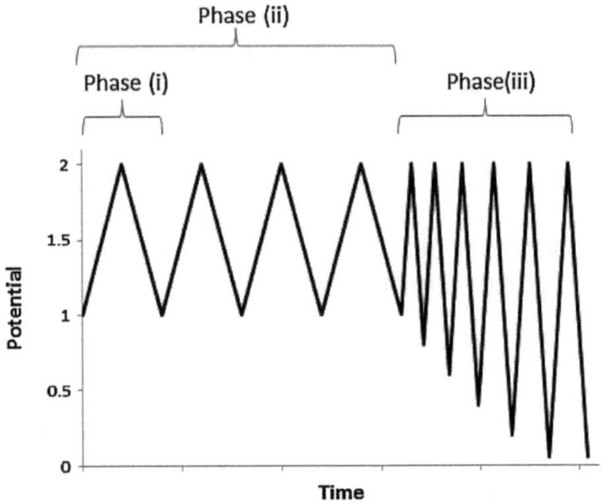

Fig. 6.1 A graph illustrating the experimental procedure used to study the electrochemical oxidation processes of the carbon black samples. The procedure aims to compare the electrochemical oxidation tendencies of the HSAC supports in three phases: (*1*) the initial oxidation (*2*) effect of repeated high potential oxidation and (*3*) effect of low potential limit

potential of the WE was then stepped to 1.05 V_{RHE} in preparation for the electrochemical oxidation study on the HSAC samples, which involved the comparison of three experimental phases. For clarity, the three phases of these potential perturbations are depicted in Fig. 6.1.

The three phases of the electrochemical treatment commenced as follows:

1. The very first high potential excursion between 1.05 and 2.05 V_{RHE} to represent the initial electrochemical oxidation of the HSAC supports at 5 mV s^{-1}.
2. A further three CVs between 1.05 and 2.05 V_{RHE} to study and compare the changes in the electrochemical oxidation tendencies of the HSAC samples at 5 mV s^{-1}.
3. The effect of the lower potential limit of the CV on the electrochemical oxidation tendencies of the now oxidised HSAC support samples by progressively lowering the lower CV potential limit between 1.05 and 0.05 V_{RHE} in steps of 200 mV maintaining an upper electrode potential limit of 2.05 V_{RHE} in six subsequent CVs at 20 mV s^{-1}.

After the electrochemical oxidation treatment the oxidised HSAC samples were then characterised once again using CV between 0.05 and 1.05 V_{RHE} at various scan rates.

6.3 Results and Discussion

6.3.1 Determination of the Apparent Double-Layer Capacitance

In electrochemistry it is often desirable to be able to quantify the mass and/or ECSA of the electrode material of interest in situ of the electrochemical half-cell experiment, which in the study of precious metal catalysts the ECSA may be relatively straightforwardly determined using a probe molecule. For example, the

ECSA of a Pt electrode may employ either $H_{ads/upd}$ or CO-Stripping (see Chaps. 4 and 5 respectively) techniques, whereas non-Pt catalysts such as Ag the ECSA can be measured via a metal-stripping (such as Pb) procedure [38–40]. However, for surfaces where a probe molecule does not adsorb (such as carbon surfaces) the in situ determination of the mass and/or ECSA is comparatively problematic with no common methodology. Consequently, a method that utilises the property of double-layer capacitance was developed as part of this study in order to quantify the mass of the carbon sample on the WE substrate using CV in situ, prior to electrochemical experimentation.

An example series of CVs for the EC300 sample black performed at a number of scan rates between 0.05 and 1.05 V_{RHE} is shown in Fig. 6.2a. An almost rectangularly shaped voltammogram is observed which appears to scale proportionally with the potential scan rate, both of which are characteristic of capacitive processes. The capacitance of the electrode surface can be extracted from the current in the CV using the following equation:

$$Q = \frac{I_{+} - I_{-}}{2v}$$

where Q = capacitive charge, I_{+} and I_{-} are the currents of the positive and negative going potential sweeps, and v = potential scan rate. The resulting apparent double-layer capacitance of the electrode may then be plotted as a function of the electrode potential as given in Fig. 6.2b. Here we observe that the capacitance does not remain constant for all potentials between 0.05 and 1.05 V_{RHE} but instead possesses a parabolic relationship with the electrode potentials during the CV (ranging from 3.6 to 3.8 mF cm^{-2}), which is furthermore slightly dependent upon the potential scan rate, converging at higher rates.

The largely featureless CV and scaling of electrode current with scan rate strongly indicate that the electrode current is dominated by capacitive currents determined by the interface between the carbon electrode and electrolyte. It is generally assumed that the whole internal structure of the carbon is wetted by the electrolyte, although this may not be strictly the case for microporous carbons [41]. The parabolic relationship between capacitance and electrode potential is in agreement with the general trend observed using CV but would ideally be constant for all potentials where no faradaic electrode reactions occur [41]. This has been quite extensively studied, and it has been noted that there may be complications with porous electrodes because of distributed capacitance effects, whereby it is speculated that the exterior of the porous carbon charges faster than the interior and that less of the surface is therefore accessible at fast scan rates [40, 41]. Nonetheless, a slight dependence of the double-layer capacitance on the CV scan rate does not impact on aims of the methodology.

Although, the minimum in the double-layer capacitance vs. potential curves (see Fig. 6.2b) are most often quoted in the literature [42] for the purpose of scaling the apparent double-layer capacitance with sample loading a value at any potential where no faradaic current is observed can be used. The apparent double-

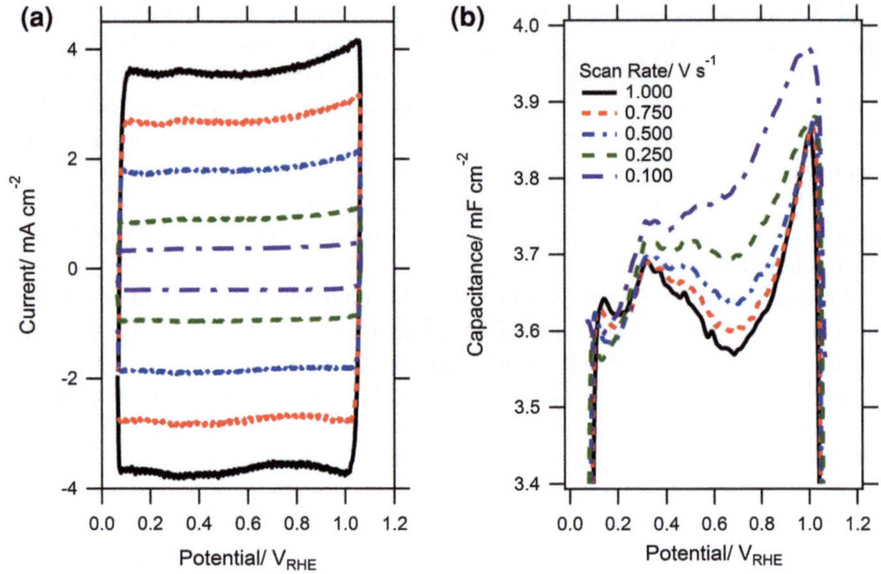

Fig. 6.2 CV of EC300 HSAC support (**a**) and corresponding potential dependence of the double layer electrode capacitance (**b**) at various potential scan rates. Measurement performed in Ar saturated 0.5 mol dm^{-3} H_2SO_4 electrolyte.

layer capacitance of the carbon samples taken at 0.25 V_{RHE} is plotted as a function of the sample loading on the WE substrate in Fig. 6.3.

In this plot we observe that the electrode capacitance scales linearly within the applied range of sample loading of each carbon black, and we can clearly see that the dependence of the capacitance of the EC300 carbon on the sample loading is significantly greater than all the heat-treated samples. Details of gravimetric and specific capacitance (BET normalised) of the carbon samples extracted from Fig. 6.3, along with details of the BET area of the carbon samples are summarized in Table 6.1.

The BET area of the EC300 carbon black ($720 m^2{}_{BET}\ g^{-1}$) is significantly greater than that of the heat-treated ($\leq 247\ m^2{}_{BET}\ g^{-1}$) samples, which decreases further with increasing graphitisation temperature, particularly at 3,200 °C ($135\ m^2{}_{BET}\ g^{-1}$). This observation is roughly mirrored by the gravimetric capacitance owing to the dependence of the double-layer capacitance on the surface area of the electrode, which in principle should (but not necessarily in practice) be rather independent of the electrode material and only on the surface area of the interface. For very porous carbons, for example, it is has been observed that the gravimetric capacitance of the porous carbon does not increase above ~90 F g_c^{-1} for carbons with specific surface areas >1,200 $m^2\ g^{-1}$ although this is far higher than the carbon samples in this study [43]. The double-layer capacitance normalised per BET area are, however, not constant ranging from between 3.4 and 8.5 μF cm^{-2} exhibiting no particular trend with heat treatment temperature. The range of capacitance values are

Fig. 6.3 Measured capacitance of the HSAC support samples plotted as a function of carbon black sample loading. The WE substrate background corrected capacitance scales linearly with the sample loading of the carbon blacks

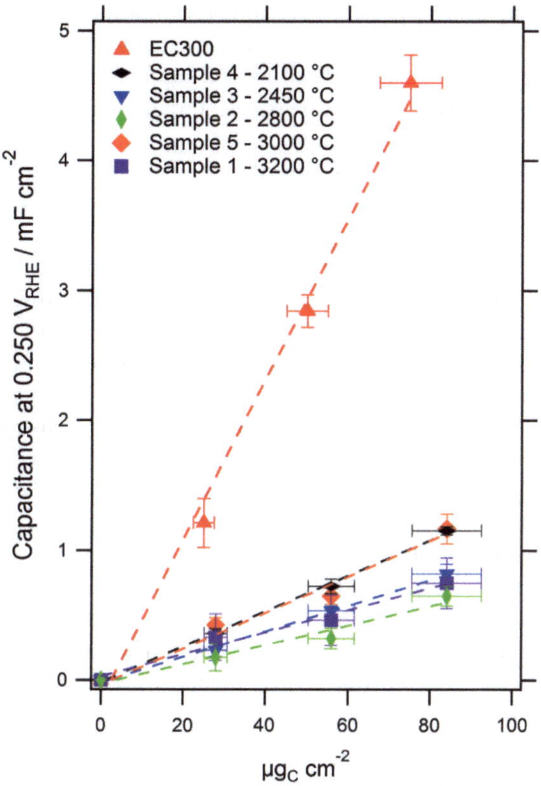

comparable to literature values, which typically lie in the range of 4 to 10 μF cm^{-2}_{BET} for a number of carbon blacks in various electrolyte [42, 43]. More specifically, the 8.5 μF $cm^{2}{}_{BET}$ observed for the EC300 carbon is closely comparable to the 8 μF cm^{-2}_{BET} of Vulcan XC-72 [42]. The sample loading can therefore be calculated using the following simple equation:

$$W_{sample} = \frac{C_{total} - C_{substrate}}{C_g}$$

where W_s = sample loading, C_{total} = total electrode capacitance, $C_{substrate}$ = WE substrate capacitance and C_g = gravimetric capacitance (pre-determined) of the carbon sample.

The significance of this methodology is that the carbon black sample loading on the WE substrate may be quantified in situ rather than relying on the deposited carbon sample remaining on the WE substrate during the course of the measurements, which in practice is prone to detaching from the WE substrate when first submerged in electrolyte reducing the actual sample loading. By measuring the sample loading in situ the experiment may be validated or adjusted to the actual loading. Furthermore, given the importance of quantifying the electrode sample size in electrocatalysis, it may also be possible to adopt a similar methodology to quantify the sample loading of non-precious metal porphyrin ring ORR catalysts. The methodology could also be extended to the study of HSAC supported Pt catalysts by pre-adsorbing a saturated CO_{ads} layer on Pt to suppress faradaic electrode processes and limiting the upper potential limit to below ca. 0.5 V_{RHE} (at avoid oxidising the CO_{ads} layer) during the CV. The electrode capacitance of the whole HASC supported Pt catalyst may be determined, similarly to a method used to assess the double-layer capacitance of an MEA in Ref. [11].

It is, however, important to note that the scaling of the gravimetric capacitance with the carbon loading, particularly for the lower surface area graphitised carbons (namely, samples with BET areas $\leq 250\ m^2\ g^{-1}$) is by no means one hundred per cent reliable and is prone to considerable error, predominantly arising from the potentially significant influence of the WE substrate background capacitance contributions. The capacitance of the WE substrate is ideally substantially smaller than the sample material and must remain constant for each experiment. The capacitance of a typical GC substrate (specific capacitance of 25 μFcm^{-2} [44]) however, was observed to be $\sim$1–2 mF cm^{-2} which is a quite considerable background equivalent to a $\sim$20–30 $\mu g_C\ cm^{-2}$ loading of EC300 and much more for the graphitised samples. A polycrystalline Au WE substrate is in comparison better suited because it only possesses a capacitance of $\sim$0.35 mF cm^{-2}. Some common sense is therefore required in the interpretation of data because changes in the WE substrate capacitance, which may arise from the leakage of electrolyte into the sides of the WE substrate or from changes in the substrate surface, can and will invalidate the double-layer capacitance measurement. A final important note is that we can only assume that the current observed is double-layer capacitance (hence the use of 'apparent') because it can also contain current contributions from reversible surface redox processes which are undefined. Although this will not influence the usefulness of this methodology for assessing the loading of a pristine carbon black sample on a WE substrate at the beginning of an electrochemical experiment, it will influence the interpretation of the specific capacitance and the changes in the apparent double-layer capacitance that can occur as a result of an electrochemical treatment, as we will be noted later in this study.

6.3.2 Substrate Background Contributions

At the beginning of the study on the electrochemical oxidation of carbon blacks, it was crucial to choose an appropriate WE substrate that (ideally) does not contribute to the faradaic electrode corrosion current, or evolve any CO_2 in the potential region of interest, here <2.05 V_{RHE}. The electrochemical oxidation of the commonly employed GC substrate, and a polycrystalline Au substrate were evaluated with comparison of the polycrystalline Pt (for reference) in Fig. 6.4.

The GC substrate exhibits rather small electrode current dominated by capacitive currents in the potential region <1.4 V_{RHE}. At higher potentials, however, the

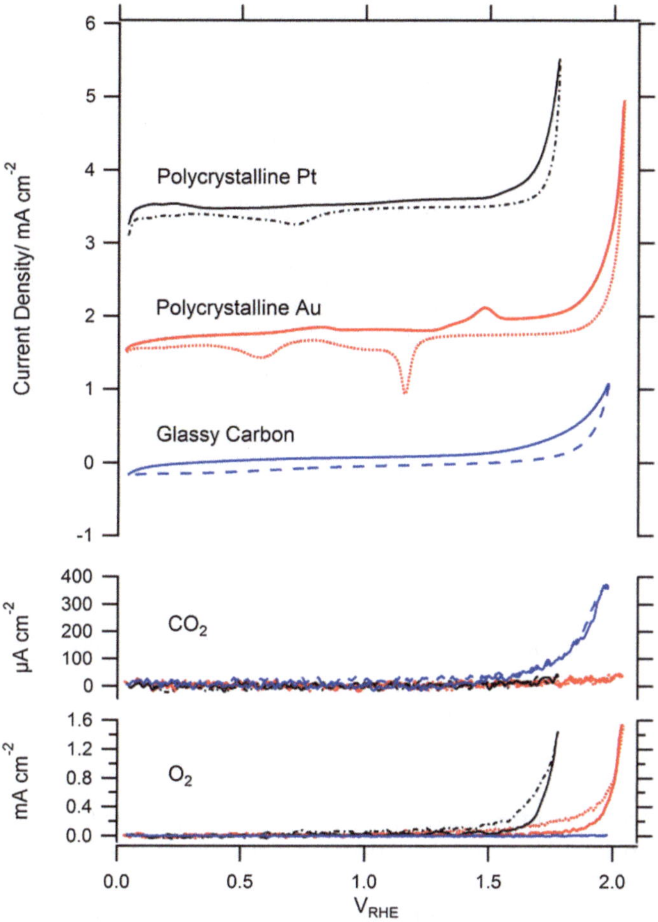

Fig. 6.4 CV and MSCVs of polycrystalline –Pt, –Au and GC electrodes. Measurements performed at 20 mV s^{-1} in 0.5 mol dm^{-3} H_2SO_4 and 300 µl min^{-1}, with corresponding CO_2 ($^{m}/_{z}$ = 44) and O_2 ($^{m}/_{z}$ = 32) mass ion currents. The commonly used GC substrate is observed to electrochemically oxidise to CO_2 at potentials above 1.5 V_{RHE} giving rise to undesirable WE substrate background contributions in the study of the electrochemical oxidation of carbon blacks

GC substrate electrochemically oxidises yielding CO_2. In comparison, the polycrystalline Au substrate exhibits pronounced features in the CV <1.5 V_{RHE} which are attributed to the oxidation and reduction of the Au surface. At potentials above 1.7 V_{RHE} the electrode current on polycrystalline Au is dominated by the OER yielding O_2. For reference, the OER on the polycrystalline Pt surface occurs at the lower potential of 1.5 V_{RHE}.

Despite that GC is widely considered as one of the most stable, chemically inert forms of carbon, the appearance of electrode current corresponding to the electrochemical oxidation of the GC to CO_2 at potentials >1.4 V_{RHE} is far from an ideal characteristic of the WE substrate background in the study of the electrochemical oxidation of HSAC supports. This effectively limits the study of the COR of HSAC supports to the region <1.4 V_{RHE}, although GC substrates have been used in previous DEMS studies on the COR [30, 31]. In comparison, a polycrystalline Au substrate does not introduce any background CO_2, evolving only O_2 at potentials >1.7 V_{RHE} allowing rather higher electrode currents to be studied. In consideration that we are primarily interested in CO_2 the potential limit can conceivably be raised above 1.7 V_{RHE}, however, in practice the maximum potential is limited by the quantity of O_2 evolved in the OER on the polycrystalline Au to <2.05 V_{RHE}. At higher potentials the rate of OER produces an excessive amount of O_2 which forms bubbles in the thin-layer and flow cell lines causing a loss of contact between the working, counter and reference electrodes. Finally, the observation of CO_2 rather than O_2 for GC is not in agreement with the claim that the oxygen evolution reaction occurs on GC electrodes in 0.1 mol dm^{-3} in $HClO_4$ occurs at 1.0 V_{SCE} (~1.35 V_{RHE}); however, these measurements were based on RDE measurements alone [45].

6.3.3 Electrochemical Oxidation of HSAC Supports

6.3.3.1 Phase (1): First High Potential Excursion

The contributions of the various electrochemical process towards the overall faradaic electrode current observed during the first potential sweep [phase (1)] from 1.05 to 2.05 V_{RHE} of the pristine carbon black samples deposited on a polycrystalline Au substrate were elucidated using DEMS. An example measurement, highlighting the relative contributions of the different electrochemical reaction processes that occur during the initial high potential excursion for the EC300 carbon >1.05 V_{RHE} is first illustrated in Fig. 6.5.

At potentials <1.2 V_{RHE} the electrode currents are dominated by double-layer capacitance currents until the potential exceeds ~1.2 V_{RHE} where we begin to observe the polycrystalline Au substrate surface oxidation peak (refer to Fig. 6.4 previously). As the potential sweeps above >1.2 V_{RHE} there is a steady increase in electrode current which is accompanied by the complete COR to CO_2. As the electrode potential is raised even further to >1.7 V_{RHE} the complete COR begins to surge in an exponential manner. Above 1.9 V_{RHE} we then observe the OER on the polycrystalline Au substrate with some evidence to indicate that a small amount of O_2 is evolved from the EC300 HSAC support (although as we will see this is specific to EC300 and is short-lived). The quantity of CO_2 and O_2 observed using the calibrated DEMS instrument was found to be clearly insufficient to account for all electrode current observed during the high potential excursion for each of the carbon samples. In this study, the remaining current is assumed to arise from the partial COR of the HSAC support which may be to surface oxide

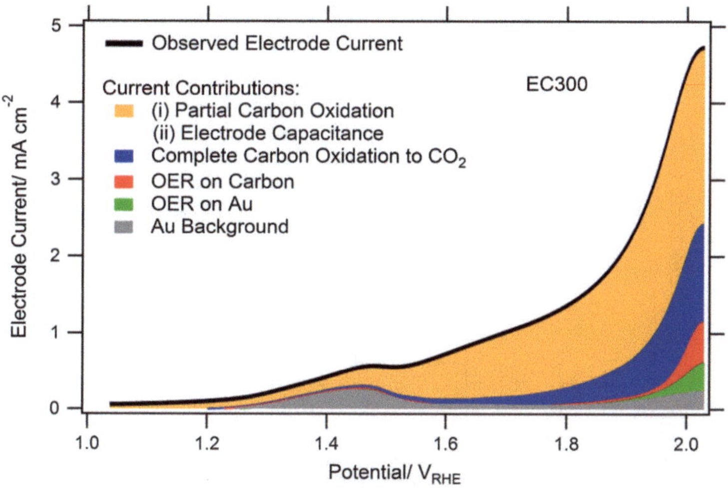

Fig. 6.5 A *graph* illustrating the origins of the observed current during an initial high potential excursion above 1.05 V_{RHE}.Pure EC300 HSAC support (50 μg_{Pt} cm^{-2}, 0.5 mol dm^{-3} H_2SO_4 and 5 mV s^{-1}). The electrode current at these high potentials is dominated by the partial and complete COR of the HSAC support

functional groups or CO, both of which are understood to occur (to varying degrees) simultaneously with the complete COR to CO_2 (see Ref. [35]).

In order to begin to compare the electrochemical oxidation tendencies of various HSACs, an initial comparison between the first high potential excursion between 1.05 and 2.05 V_{RHE} for each of the carbon black samples is given in Fig. 6.6. It is immediately clear that the electrode current observed on the EC300 above 1.0 V_{RHE} is significantly greater than for graphitised samples (despite that the EC300 mass loading is almost half in this figure i.e. 50 vs. 85 μg_c cm^{-2}). For all the carbon samples, the observed electrode current coincides with the COR to CO_2, exhibiting a clear trend for decreasing COR rates with increasing graphitisation temperature. The peak observed at 1.45 V_{RHE} and the O_2 observed >1.7 V_{RHE} in each of the measurements can be solely attributed to the OER on the polycrystalline Au substrate, with the exception of the EC300 sample where some additional O_2 was observed in the initial excursion. Finally, the overlap of the O_2 evolved in the OER on the polycrystalline Au substrate for heat-treated sample demonstrate the precision and reproducibility of the DEMS measurement.

In the example Fig. 6.5, it was highlighted that the COR of the EC300 HSAC support to CO_2 coincided with additional electrode current that was attributed to partial COR processes. By subtracting the I_{CO2}, I_{O2} and Au WE substrate background current contributions measured using DEMS from the overall observed electrode current, the currents attributed to the partial oxidation of the carbon black can be deduced. The integrated currents corresponding to the charge arising from the complete and partial COR on the carbon black samples during the first

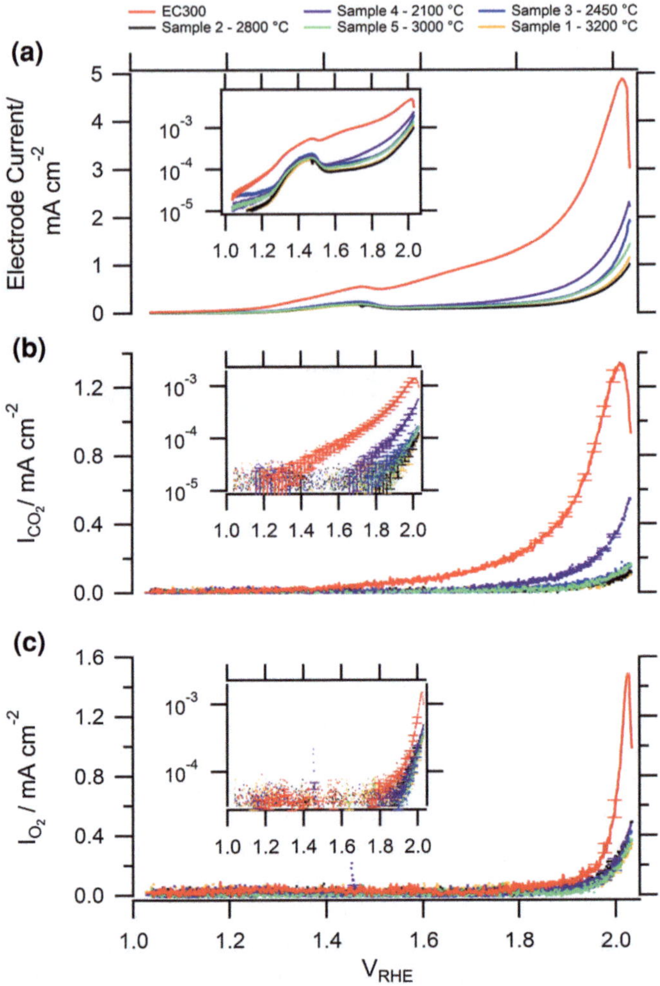

Fig. 6.6 First high potential LSV (5 mV s^{-1}) of the pristine EC300 (50 μg_c cm^{-2}) and heat-treated samples 1–5 (85 μg cm^{-2})]. In Argon saturated 0.5 mol dm^{-3} H_2SO_4 electrolyte on a polycrystalline Au substrate (**a**) I_{CO2} (**b**) and I_{O2} (**c**). The corresponding log I vs. V_{RHE} plots are inset. The rate of electrochemical oxidation of the HSAC support is comparably high for the pristine EC300 sample with an observed trend for decreasing COR rate with increasing graphitisation temperature

potential excursion to 2.05 V_{RHE} are summarised as a function of the heat-treatment temperature in Fig. 6.7.

In this plot we observe that the mass normalised charge for the complete and partial COR of the carbon blacks in general decreases with increasing graphitisation temperature. The complete COR charge evolved by the EC300 is almost one order of magnitude greater than the graphitised samples treated ≥2,450° C. However, the charge corresponding to the partial oxidation of the EC300 HSAC

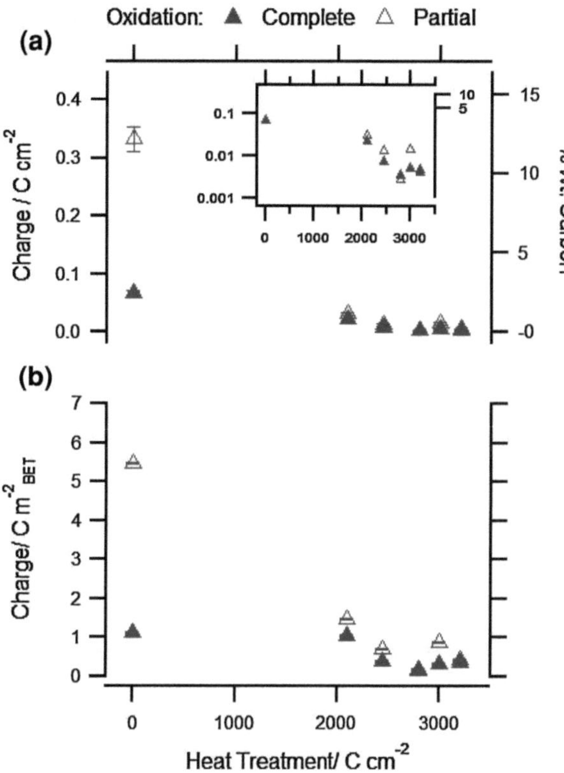

Fig. 6.7 Effect of heat treatment temperature on the complete and partial electrochemical oxidation charge evolved during the first high potential excursion to 2.0 V_{RHE} (**a**) (EC300 normalised to 85 μg cm^{-2}).Corresponding oxidation current normalised to BET surface area (**b**) in 0.5 mol dm^{-3} H_2SO_4 for the various HSAC samples. The quantity of partial COR is significantly higher than complete COR on the EC300 sample. The amount of partial and complete COR is highest on the EC300 HSAC support. The mass normalised quantity of complete electrochemical COR of the EC300 is, however, normalised to BET are the values are much more comparable.

support is significantly larger than complete COR in contrast to the heat-treated samples where the charges are of similar magnitude. Finally, by normalising the COR charges to the BET surface area of the HSAC samples it is observed that the specific complete COR charge evolved from the EC300 and graphitised sample 4 (2,100° C) are quite similar.

In the logarithmic current vs. potential plots shown in the inset of Fig. 6.6, a linear Tafel-slope region for the I_{CO2} and I_{O2} can be observed in the oxidation of each of the HSAC samples, although in general these regions cover less than one decade of current. The CO_2 detection limit potential, Tafel-slope and corresponding transfer coefficient (α*n) extracted from the observed I_{CO2} are summarised in Table 6.2.

In this table we precisely see that the CO_2 detection limit potential for the complete COR of the carbon blacks increases with increasing heat-treatment temperature, with clear potential steps between the EC300, sample 4 (2,100° C), and the remaining graphitised samples (>2,450° C). For the EC300 and sample 4 (2,100 °C) carbons, two Tafel-slope linearities are observed in the logarithmic plot above and below 1.8 V_{RHE}. Below 1.8 V_{RHE} the slopes are comparatively shallow ($\geq$307 mV Dec^{-1}); however, above 1.8 V_{RHE} the Tafel-slopes for all the carbon

Table 6.2 COR Tafel slopes (I_{CO2}) and corresponding symmetry/transfer coefficient constants for the complete oxidation of each HSAC during the first high potential excursion

Sample	Graphitisation temp./ °C	CO_2 detection limit potential/ V_{RHE}	Tafel slope/ mV Dec^{-1}	α*n
EC300	–	1.35	*413 ± 6	0.143
			229 ± 2	0.258
(4)	2100	1.68	*307 ± 11	0.192
			170 ± 2	0.348
(3)	2450	1.84	200 ± 9	0.295
(2)	2800	1.88	206 ± 11	0.287
(5)	3000	1.88	167 ± 6	0.354
(1)	3200	1.93	170 ± 9	0.348

DEMS I_{CO2} detection limit equivalent to 25 μA cm^{-2} . In cases where two linear regions were observed, two Tafel slopes are given, whereby the * avalues corresponds to the slope in the lower potential region <1.8 V_{RHE}. The α*n obtained from these slopes are comparable to values used in kinetic models of the HSAC support degradation in PEMFCs.

black samples are comparable (~200 ± 30 mV Dec^{-1}) with no clear relationship with the heat-treatment temperature.

Discussion of Initial Electrochemical Oxidation Processes

It is uncommon to begin the discussion of experimental data with the limitations of the study, however, it is necessary in this case in order to focus the discussion on how the DEMS instrument can be used to study the COR, rather than overly speculate on the possible COR mechanism(s) that occur during each phase of the electrochemical treatment. The obvious limitation here is that we simply do not know enough about the bulk and surface characteristics of the pristine carbon black materials i.e. BET surface areas alone are not enough. For the discussion in this study we instead have to make the following (reasonable) assumptions concerning the relative bulk and surface characteristics of the pristine EC300 and graphitised samples: (1) the EC300 carbon black possesses the least graphitic character and the highest oxygen content, whereas (2) the heat-treated samples possess more graphitic character, which increases with rising heat-treatment temperature that in turn, decreases the BET area and oxygen content. These assumptions are based on previous observations that the oxygen content is in general proportional to the BET surface area [21] (although heat-treated samples tend to have lower oxygen content) and that heat-treatment organises the graphene layers reducing the amorphous nature of the carbon decreasing the d-spacing (and decreasing BET area) [46]. For more in-depth discussion of the differences in the initial COR tendencies and their mechanism(s), a greater effort in the characterisation of the pristine carbon samples using a range of methods (along with BET) is required. For example, techniques such as X-ray diffraction (XRD) can be used to determine the interlayer d-spacing and deduce the graphitic character of the carbon [21, 46], whereas *iR*-spectroscopy [47] or XPS [26] may be used to identify the nature and (possibly) concentration of the surface oxygen containing functional groups that are initially present on the pristine carbon samples.

In phase (1) of the electrochemical treatment, DEMS was able to elucidate the proportion of electrode current arising from the complete COR to CO_2, and the OER on polycrystalline Au evolving O_2 during the high potential excursion to 2.05 V_{RHE}. By subtracting the calibrated I_{CO2} and I_{O2} observed using DEMS, the remaining electrode current attributed to the partial COR can be deduced, an assumption that the complete COR occurs in parallel with the partial COR of carbon blacks, based on the findings of previous studies [33]. The partial COR current was observed to be quite significant, contributing toward ~70 % of the COR charge for the pristine EC300 and in general ~50 % for the graphitised samples during the first potential excursion to 2.05 V_{RHE} and may proceed to both oxygen containing surface functional groups and CO [18, 33], the latter of which of course could also conceivably elucidated using DEMS. In light of the extensive partial COR current observed, the in situ observation of CO would certainly be worthwhile measurement using DEMS to evaluate the quantity of CO in the future, although not necessarily straightforward and is discussed in more detail later in Sect. 6.4.6. Nonetheless, the significant partial COR current observed would suggest that it should not be assumed that the whole electrochemical oxidation current arises from the complete COR of carbon black to CO_2 and necessarily leads to carbon weight loss.

The lower complete and partial COR rates observed for the graphitised carbon black samples (with a tendency to decrease with increasing heat-treatment temperature) in comparison to the pristine EC300 is in agreement with the well-established understanding that the heat-treatment and consequent graphitisation of carbon blacks tend to increase their resistance to chemical and electrochemical oxidation [35]. These finding are, for instance, in agreement with a previous DEMS study on an unspecified carbon black, whereby the CO_2 evolved from a heat-treated material was found to be lower than that of the pristine sample 1.5 V_{RHE} [30]. In an electrochemical half-cell study performed at 1.2 V_{RHE} and 80 °C, the corrosion current of non-graphitised carbons were observed to be in general, proportional to their BET area; however, graphitised carbons performed better by factor of ~5, in general agreement with the findings for the BET normalised partial and complete COR charges for the samples heat-treated ≥2450 °C [20]. In the PEMFC environment meanwhile, graphitised carbon MEAs have been found to perform up to 35 times better than non-graphitised carbon blacks at 1.2 V_{RHE}, and ~5 times better during start-stop corrosion cycles in the fuel cell environment [7].

Although the complete and partial COR rates of the graphitised samples were all found to be slower than the pristine EC300 carbon black, there appears to be a more significant decrease in the COR rates when increasing the heat-treatment temperature from 2,100 to ≥2,450° C. This would suggest that the changes in the microstructure and type and/or concentration of oxygen containing surface functional groups during the graphitisation of this particular carbon black at temperatures of 2,450 °C and above, is more desirable in terms of achieving greater electrochemical oxidation corrosion resistance (at least initially), with a trend for increasing resistance with higher heat-treatment temperatures. This is in agreement with the trend observed in Ref. [48] whereby heat-treatment at 2,500° C appeared to significantly decrease the specific COR rate (normalised to BET area) of the carbon black compared to lower temperature heat-treatment at 2,000° C. The significantly greater COR rates on the EC300 is likely be attributed to the higher concentration the electrochemical oxidisable oxygen-containing surface functional groups that are initially present on the pristine HSAC support, which can be considered as reactive centres for the COR processes [21]. Ex-situ characterisation of the bulk and surface characteristics of the pristine carbon samples, however, is needed in order to shed more light on the differences between the initial oxidation rates of the pristine EC300 and heat-treated carbon samples, without which the rationalisation is purely speculative.

The pristine EC300 carbon exhibited two features that are not observed in the graphitised samples: (1) significant partial COR current representing ~70 % of the total charge (compared to ~50 % for the graphitised samples), and (2) some evidence of O_2 evolution that is not attributed to the polycrystalline Au WE substrate. The significant proportion of partial COR contributions to the overall current is in agreement with previous findings that the proportion of partial COR dominates the complete COR for pristine carbon blacks (in contrast to their graphitised counterparts) during the initial electrochemical oxidation, but decreases with time so that the complete COR eventually becomes dominant [49]. The origin of the observed O_2 in the initial potential excursion is meanwhile unclear, and rather unexpected although the OER is thought to occur on certain carbon systems [45] (although the measurement on GC in Fig. 6.4 questions this). In phases (2) and (3) of the treatment, however, we will later observe that O_2 is specific to the initial potential sweep and is not observed in any of the subsequent potential excursions. If we consider that the O_2 feature coincides with a rapid increase in the COR rate, the O_2 evolved may have been non-oxidatively released from sub-surface voids which opened as a result of the complete COR (rather than the oxidation of water), although this speculation.

Using DEMS we were also able to extract kinetic information on the complete COR by extracting the I_{CO2} Tafel-slope and the potential at which CO_2 was first observed. In Table 6.2 this potential was carefully termed the CO_2 'detection limit potential' rather than 'onset potential' because the complete COR can occur at lower potentials but at rates that are simply below the detection limit of the DEMS instrument. No CO_2 below 1.35 V_{RHE} was observed on the EC300 carbon, 1.68 V_{RHE} on the mildly graphitised carbon (2,100° C), or 1.84 V_{RHE} on the more heavily graphitised carbons (≥2,450° C) at a detection limit that is equivalent to a carbon weight loss of 2.754 $\mu g_c\ h^{-1}$ evolved via the complete electrochemical COR, or 25 $\mu A\ cm^{-2}$ of faradaic current (assuming 4 e^- per CO_2). Previous DEMS studies have however, observed CO_2 at potentials as low as 0.9 V_{RHE} on NORIT BRX (BET of ~1,800 $m^2\ g^{-1}$ [50])[29], Vulcan XC-72 (BET of ~238 $m^2\ g^{-1}$[51]) [31, 52] and an unspecified porous carbon with high oxygen content [30]. This may suggest that the EC300 (BET of ~720 $m^2\ g^{-1}$) carbon is more resistant to the complete COR than Vulcan XC-72 (a commonly employed PEMFC catalyst support) despite its higher BET area; however, the comparison between DEMS studies is not necessarily straightforward because previous studies were not quantitative, and the CO_2 detection limit of the DEMS instruments are not given. The sample loadings of carbon material in the two more recent DEMS studies of 61 μg_c cm^{-2} [52], 142 μg_c cm^{-2} [30] are nevertheless comparable to this study (50 and 85 μg_c cm^{-2}) although a much higher loading of 5 mg_c cm^{-2} was used in the earliest DEMS study [29] which could therefore prove to be more sensitive. Possible approaches to improving the DEMS instrument sensitivity are later discussed in Sect. 6.4.6.

The Tafel-slopes extracted from the observed I_{CO2} meanwhile, were in the range of 200 ± 30 mV Dec^{-1} above 1.8 V_{RHE}, although shallower slopes of 413 and 307 mV Dec^{-1} were observed on the EC300 and mildly graphitised HSAC in the potential region below 1.8 V_{RHE}. These slopes are slightly steeper than the 270 mV Dec^{-1} obtained from the measured faradaic electrode current using chronoamperometry in phosphoric acid at the lower potential 0.8 and 1.2 V_{SHE} between 135 and 148° C for Vulcan XC-72 and its graphitised counterpart heat treated at 2,700° C [53] (and similarly Ref. [54]). The corresponding transfer coefficients (α*n) obtained from the Tafel-slopes meanwhile can be compared to values used in various PEMFC corrosion studies such 0.275 [3], 0.66 [34] 1.0 [4] which measured and/or modelled the carbon corrosion over a range of temperature, potential and time. The values obtained in the electrochemical cell environment of this DEMS study, ranged between 0.143 and 0.354, which are more comparable to the value of 0.275 used in Ref. [3]. The usefulness of the Tafel-slopes and their kinetic constants is, however, not entirely clear at this stage. It may be tempting to extrapolate the Tafel-slopes down in potential to obtain COR rates at more practical PEMFC potentials ca. <1.2 V_{RHE} (with considerable errors) however, this would assume that the potential dependence of the COR processes are rather constant over time, and that the COR rate is independent of the degree of previous oxidation, neither of which will be shown to be the case.

Overall, phase (1) of the electrochemical treatment has established that the complete COR to CO_2 occurs in combination with a significant amount of partial COR, particularly on the pristine EC300 carbon black. The electrochemical COR rates are found to be significantly lower on the graphitised samples, particularly when heat treated ≥2,450 °C with a trend of decreasing COR rates with increasing heat-treatment temperature observed. In order to speculate further on the COR mechanism(s) the HSAC supports require greater characterisation using techniques such as XRD and *iR*-spectroscopy or XPS. The observations so far, however, are only relevant to the COR processes in the very first high potential excursion.

6.3.3.2 Phase (2): Repeated High Potential Excursions

A further three high potential CVs were performed between 1.05 and 2.05 V_{RHE} in phase (2) of the electrochemical treatment in order to assess the changes that occur in the COR tendencies of each of the carbon black samples. The observed behaviours can be separated into three trends represented by the EC300, and heat treated samples 5 (3,000° C) and 4 (2,100° C), given in Figs. 6.8, 6.9 and 6.10 respectively. The data for the remaining samples are given in Appendix B.

In Fig. 6.8 we observe the first trend, which was that following the very first positive high potential excursion the electrochemical COR of the EC300 support slowed significantly in the reverse potential sweep, coming to rather abrupt end.

In the three subsequent LSVs between 1.05 and 2.05 V_{RHE}, very little or no CO_2 was observed and only small amounts of electrode current are present above the Au substrate background (see logarithmic plot inset). Furthermore, there is no evidence of the O_2 evolution from the EC300 following the first CV, whereby the O_2 measured in the three subsequent CVs can be solely attributed to the OER on the polycrystalline Au WE substrate.

The second trend, exhibited by the more heavily graphitised samples 3 (2,450 °C), 2 (2,800° C), 5 (3,000° C) and 1 (3,200° C), was that the electrochemical oxidation rate gradually increased for each successive CV, which is represented by the CVs of sample 5 (3,000° C) in Fig. 6.9.

The electrode current and corresponding I_{CO2} clearly increase with each successive LSV whilst the Tafel-slopes of the I_{CO2} (see insert) appears to remain fairly constant ($\sim$167 mV Dec^{-1}). All O_2 observed can meanwhile be attributed to the OER polycrystalline Au WE substrate background only.

The final trend demonstrated by sample 4 (2,100° C), was that the rate of electrochemical oxidation increased during the first few LSVs (1–3), but then decreased in the final fourth LSV of the phase (2) electrochemical treatment, as shown in Fig. 6.10.

The small changes observed in the measured O_2, however, reflect the quality of this particular measurement rather than any O_2 evolution on the carbon. Again, the Tafel-slopes remain reasonably constant (see inset).

In order to reveal these qualitative trends more quantitatively, the electrode charge evolved which can attributed to the complete and partial COR processes during each CV are plotted as a function of the cumulative charge corresponding to complete COR (obtained from I_{CO2}). By viewing the data in this way we are able to bring greater focus onto the dependence of the partial and complete COR processes on the prior quantity of complete COR of the carbon black sample, and therefore the significance of each CV. The data obtained during the first four CVs (whereby each point corresponds to the charge in each CV) for the carbon samples is summarised in Fig. 6.11.

In these plots, we can more clearly observe the three trends exhibited by the carbon blacks interpreted from the LSVs (in previous figures) in a more quantitative manner. In all cases, we see that the complete COR occurs simultaneously with the additional electrode charge attributed to the partial COR process. For the EC300, the faradaic charge arising from the partial COR significantly dominates that of complete COR to CO_2 in the first CV whereas for heat-treated samples the relative charges are far more comparable. In the case of the mildly graphitised carbon blacksample4 (2,100° C) the charge arising from the COR reached a maximum during the third CV. The complete and partial COR of the remaining

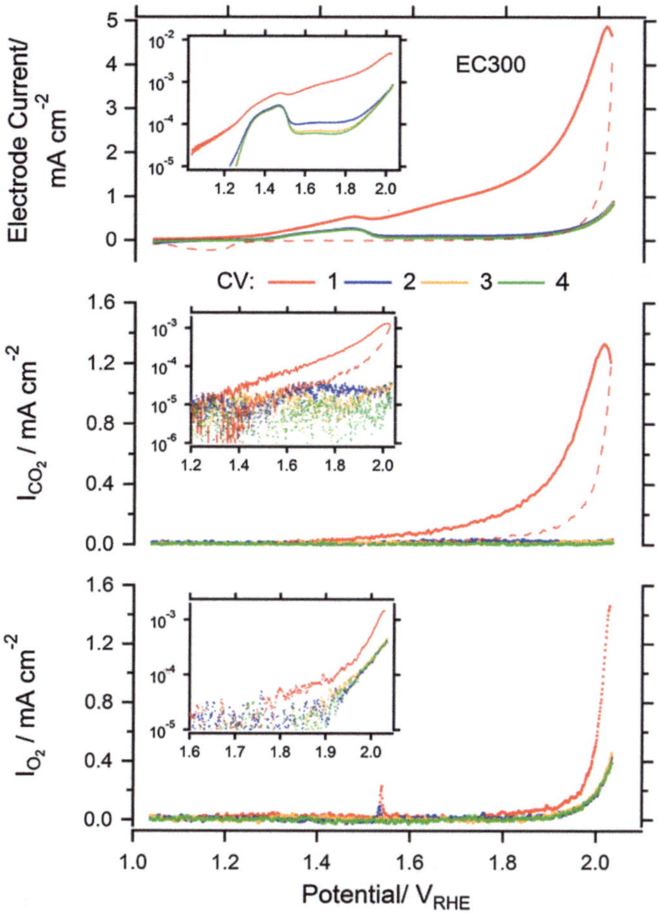

Fig. 6.8 First four LSVs of the electrochemical COR of a pristine EC300 HSAC support in 0.5 mol dm^{-3} H_2SO_4. EC300 sample loading of 50 ug_c cm^{-2} on polycrystalline Au WE substrate. Following a rather high amount of COR in the first CV, the COR ceases almost completely in subsequent CVs between 1.05 and 2.05 V_{RHE}

more graphitised HSAC samples (≥2,450° C) meanwhile increase with progressive oxidation. At this point in the electrochemical oxidation treatment, the EC300 and graphitised samples 4 (2,100° C) and 3 (2,450° C) have undergone the most complete COR to CO_2 and consequently lost the most significant weight of carbon (~3, 8 and 4.5 % respectively), although the COR have either ceased (EC300) or show signs of slowing (Sample 4). The more heavily graphitised samples (heat-treated ≥2,800 °C) have in comparison proven to be far more resistant to the complete COR and lost significant less weight in carbon (<2.5 %), although the complete COR rates are in those cases clearly increasing.

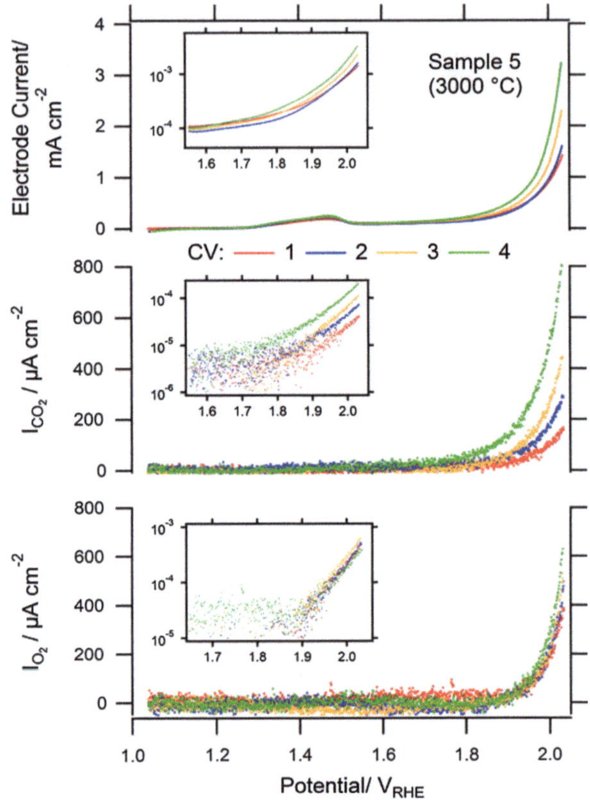

Fig. 6.9 First four LSVs of the electrochemical COR of graphitised Sample 5 (3,000° C) HSAC support in 0.5 mol dm^{-3} H_2SO_4. Sample 1 loading of 85 ug$_c$ cm^{-2} on polycrystalline Au WE substrate. The rate of electrochemical oxidation of the highly graphitised support increases with each subsequent CV between 1.05 and 2.05 V_{RHE}

In the first four CVs [phase (1) and (2)] the electrode potential was scanned between 1.05 and 2.05 V_{RHE}, and the potential of the carbon black is therefore always in thermodynamically favoured carbon oxidation potential region i.e. $\gg$0.207 V_{RHE}. Consequently, the next step of the electrochemical treatment was to investigate the effect of lowering electrode potential limit, and therefore the possible influence of the reduction on the now oxidised [in phase (1) and (2)] carbon blacks in phase (3) of the treatment.

6.3.3.3 Phase (3): Reduction of the Lower Potential Limit

In phase (3) of the electrochemical treatment, the effect of the lower electrode potential limit on the COR of the oxidised [in phase (2)] carbon blacks was investigated by progressively decreasing the lower potential limit from the 1.05 V_{RHE} [lower limit in phase (2)], to 0.05 V_{RHE} in steps of 200 mV in each of the six subsequent CVs. The observations made can be again be categorised into three trends, influenced by the consequence of the lower potential limit, each of which is represented by the EC300 and heat-treated Samples 1 (3,200° C) and 3

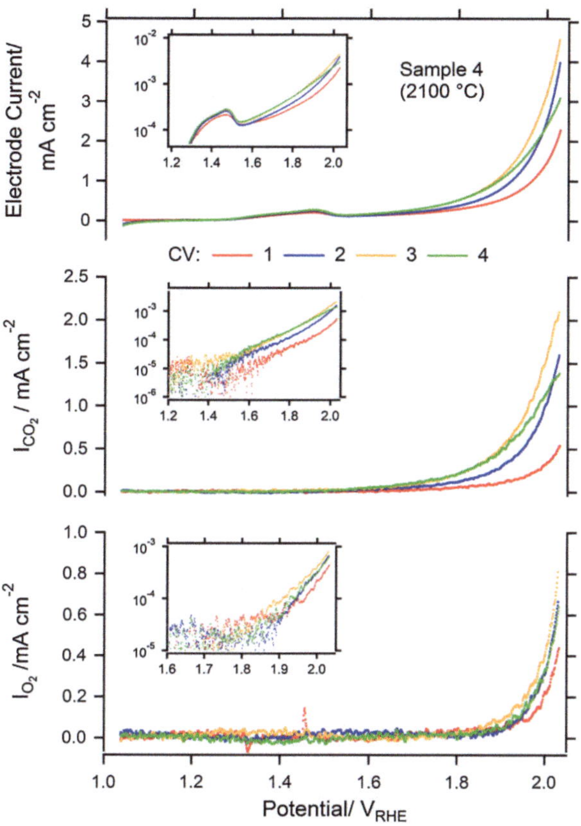

Fig. 6.10 First four LSVs of the electrochemical COR of graphitised Sample 4 (2,100 °C) HSAC support in 0.5 mol dm^{-3} H_2SO_4. Sample 4 loading of 85 ug_c cm^{-2} on polycrystalline Au WE substrate. The rate of electrochemical oxidation of the mildly graphitised support initially increases in the first three successive CVs, before being observed to decrease in the final CV between 1.05 and 2.05 V_{RHE}

(2,450° C) in Figs. 6.12, 6.13 and 6.14 respectively. The corresponding data for the remaining samples are given in Appendix B. The first trend, exhibited by the EC300 sample is shown in Fig. 6.12 . By lowering the WE potential to 0.05 V_{RHE} and reducing the EC300 HSAC, we observe that the quantity of electrochemical oxidation current that occurs during the subsequent positive high potential sweep is not only significant, but also begins at the considerably lower electrode potential of ~ 0.5 V_{RHE} peaking at ~ 1.7 V_{RHE}. Furthermore, the reduction of the HSAC appears to coincide with some non-oxidative evolution of CO_2 at potentials$<$0.207 V_{RHE}.

The second electrochemical oxidation trend was exhibited by the HSAC carbon samples heat treated at the higher temperatures ≥2,800° C, and is demonstrated by sample 1 (3,000° C) in Fig. 6.13.

In contrast to the profound effect that the reduction of the EC300 carbon black had on the complete COR to CO_2, these heat-treated samples are seemingly unaffected with no oxidation peak at 1.7 V_{RHE}. Instead the samples follow the same increasing trend in the COR rates observed in phase (2) of the experiment.

The final trend is observed for both the mildly heat-treated samples 4 (2,100° C) and 3 (2,450° C), the latter of which is given in Fig. 6.14.

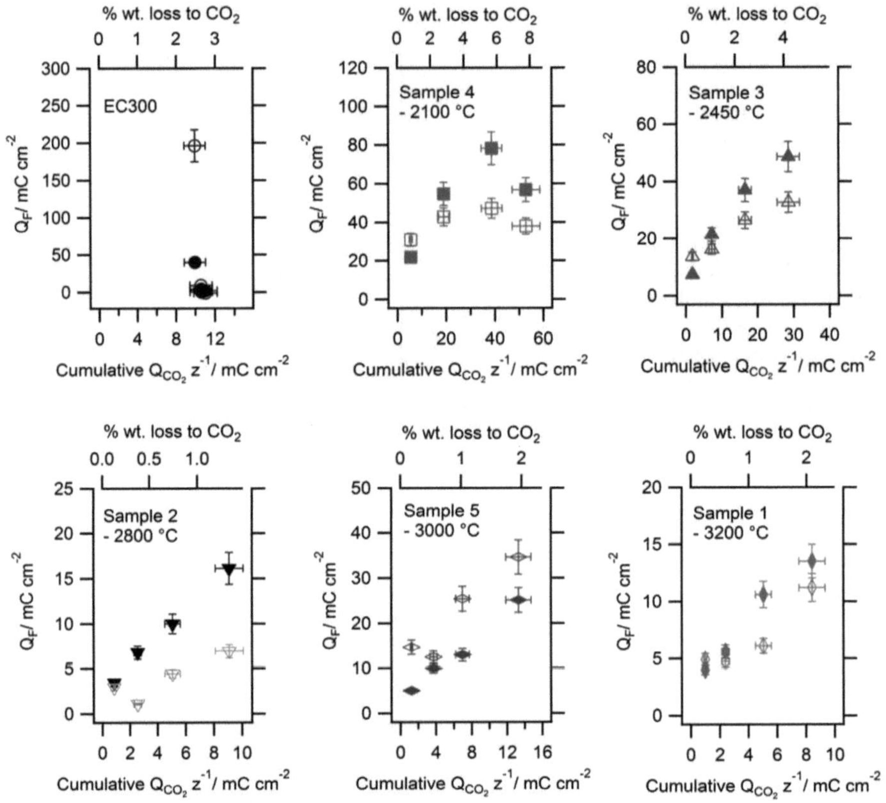

Fig. 6.11 Integrated charge of the partial (*hollow markers*) and complete (*solid markers*) electrochemical COR processes for each of the carbon black samples during the first four CVs between 1.05 and 2.05 V_{RHE} in 0.5 mol dm^{-3}. Three trends are observed, either the quantity of electrochemical COR is: (*1*) initially high and ends abruptly (EC300), (*2*) initially increases and then falls (Sample 4) or (*3*) increases (Samples 1–3 and 5), for each successive CV

By lowering the WE potential to 0.05 V_{RHE} and reducing the carbon surface, a COR peak at ~1.7 V_{RHE} along with a hint of the non-oxidative release of CO_2 <0.207 V_{RHE} which coincides with reduction processes of the carbon black, as was observed on the EC300 sample. In this case, however, the electrochemical oxidation of the mildly graphitised carbon blacks had not ceased during the first four CVs (although sample 4 was decreasing) in phase (2) of the treatment, and consequently the COR is enhanced rather than re-initiated. In all of the measurements during phase (2 and 3) the O_2 evolved can again be attributed to the OER on the polycrystalline Au substrate.

In order to once again, summarise the trends observed in the phase (3) of the electrochemical treatment, a plot of the partial and complete electrochemical COR charges is given as a functional of the cumulative CO_2 evolved for each CV in Fig. 6.15. In the interpretation of this summary plot, it should be kept in mind that

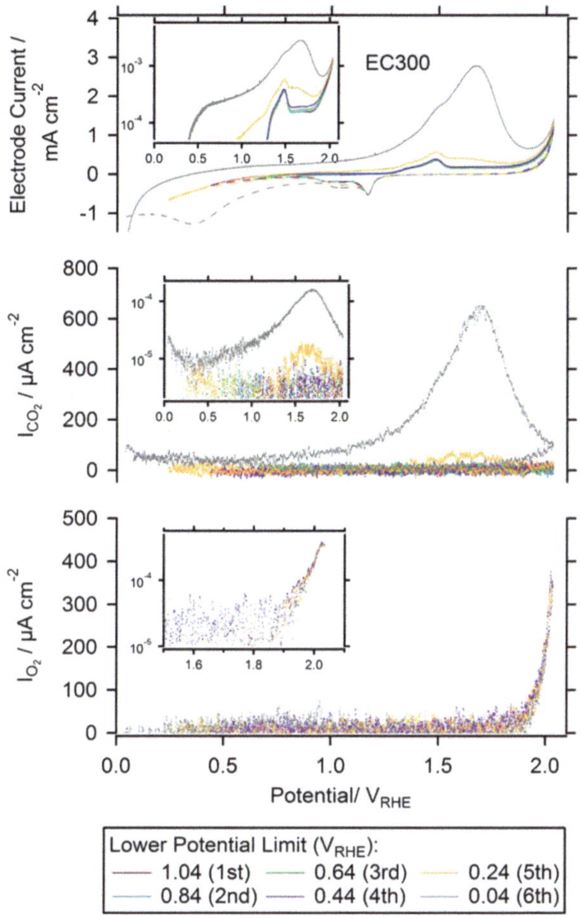

Fig. 6.12 Effect of the lower potential limit on the electrochemical COR of electrochemically oxidised EC300 carbon black in 0.5 mol dm^{-3} H_2SO_4. Sample loading of 50 ug$_c$ cm^{-2} on a polycrystalline Au WE substrate. The electrochemical COR is significantly re-activated by reducing the lower electrode potential limit to <0.25 V_{RHE} of the electrochemically oxidised EC300 sample

for each point, corresponding to each CV, the lower limit was progressively decreased by 200 mV, from 1.05 to 0.05 V_{RHE}.

In this figure we observe that both the partial and complete COR of the carbon surface occur simultaneously on all the carbon blacks [analogous to phase (2)]. For the EC300 and mildly graphitised samples [samples 3 (2,450° C) and 4 (2,100° C)], the charge corresponding to the partial and complete oxidation is increased significantly upon lowering the electrode potential to 0.05 V_{RHE}. In contrast, there appears to be little or no effect on the electrochemical oxidation of the carbon samples treated at the higher temperatures (≥2,800° C), although it is difficult in this plot to interpret with any certainty. In order to bring greater focus to the influence of the lower electrode potential limit therefore, the differential complete COR charge for each subsequent CV is plotted as a function of the lower electrode potential limit in Fig. 6.16.

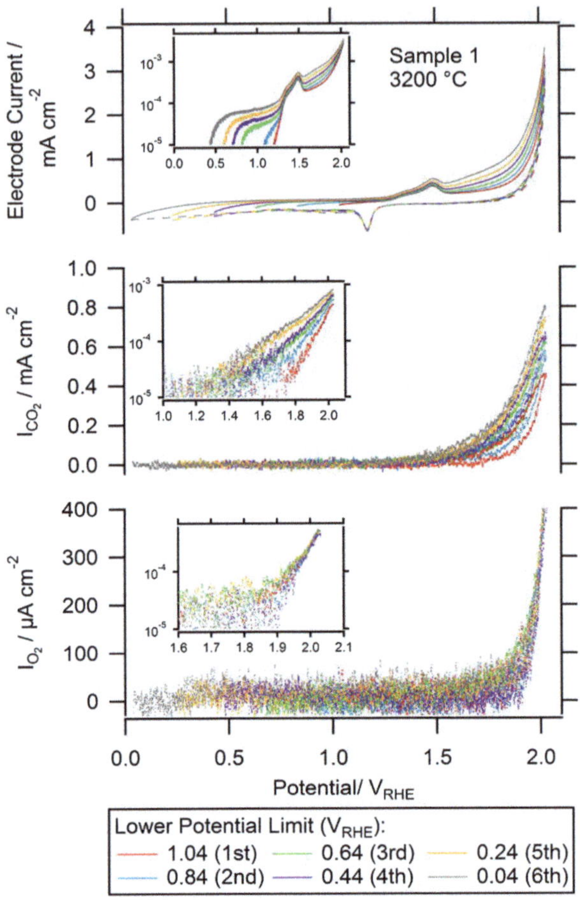

Fig. 6.13 Effect of the lower potential limit on the electrochemical COR of electrochemically oxidised and graphitised Sample 1 (3,200° C) carbon black in 0.5 mol dm^{-3} H_2SO_4. Sample loading of 85 ug$_c$ cm^{-2} on a polycrystalline Au WE substrate. The lower electrode potential limit does not appear to influence the electrochemical COR of the partially electrochemically oxidised highly graphitised HSAC sample

This plot clearly shows that by decreasing the lower electrode potential limit to <0.44 V_{RHE}, the complete COR charge increases in the subsequent CV on the EC300 and mildly heat-treated (2,100–2,450° C) samples which exhibited the peak at 1.7 V_{RHE}. In contrast, the electrochemical oxidation COR processes of the samples treated at higher temperatures (>2,800° C) remain uninfluenced by the lowering of the electrode potential limit.

6.3.3.4 Discussion of the Long-term COR Tendencies

Phases (2) and (3) of the electrochemical oxidation treatment demonstrated that the COR processes are dynamic, revealing three trends in the behaviour of the carbon black samples to repeated electrochemical oxidation, which may furthermore be influenced by the reduction of the carbon. Despite the initially high rates of the complete and (in particular) partial COR on the EC300, the oxidation of the

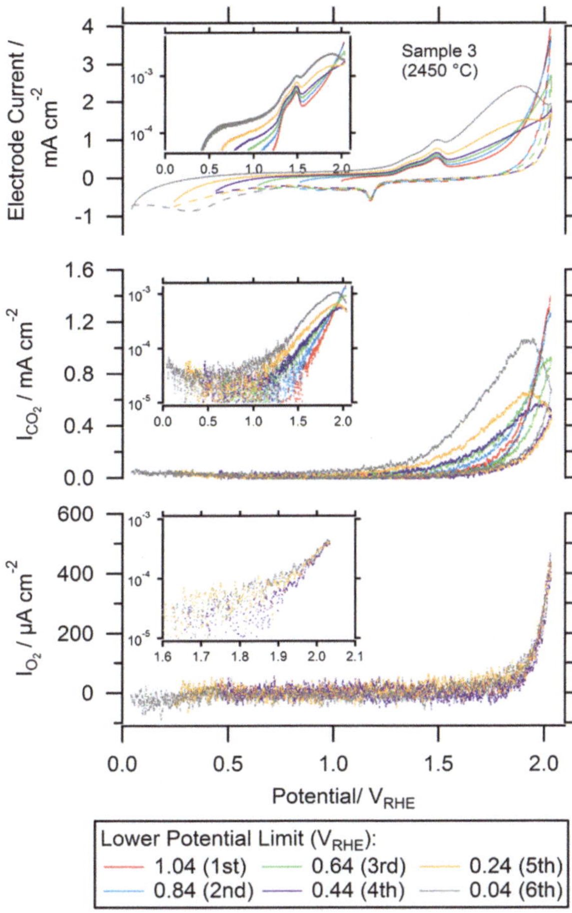

Fig. 6.14 Effect of the lower potential limit on the electrochemical COR of electrochemically oxidised and graphitised Sample 3 (2,450° C) carbon black in 0.5 mol dm^{-3} H_2SO_4. Sample loading of 85 ug$_c$ cm^{-2} on a polycrystalline Au WE substrate. The electrochemical COR is re–activated by reducing the lower electrode potential limit to <0.25 V_{RHE} of the partially electrochemically oxidised mildly graphitised HSAC sample

carbon surface was observed to be comparatively short-lived. In contrast, the COR rates on the more heavily graphitised samples (heat-treated ≥2,800° C) progressively increased during phase (2) of the treatment whereas in the final trend exhibited by the mildly graphitised samples (≤2,450° C), the COR rate initially increased before showing signs of decreasing in phase (2) or early in phase (3) of the electrochemical treatment. In phase (3) of the experiment, it was then found that by decreasing the lower electrode potential limit to <0.240 V_{RHE} and reducing the carbon, the complete and partial COR rates of the EC300 and mildly graphitised carbon blacks could be re-initiated or enhanced, giving rise to a new COR feature peaking at 1.7 V_{RHE}. In stark contrast, the COR processes of the heavily graphitised carbons were seemingly unaffected by the lower electrode potential limit, simply continuing the trend observed in phase (2) of the electrochemical treatment. Essentially, the COR rates can progressively decrease or increase, and by reducing the carbon, become re-initiated or enhanced in subsequent oxidation.

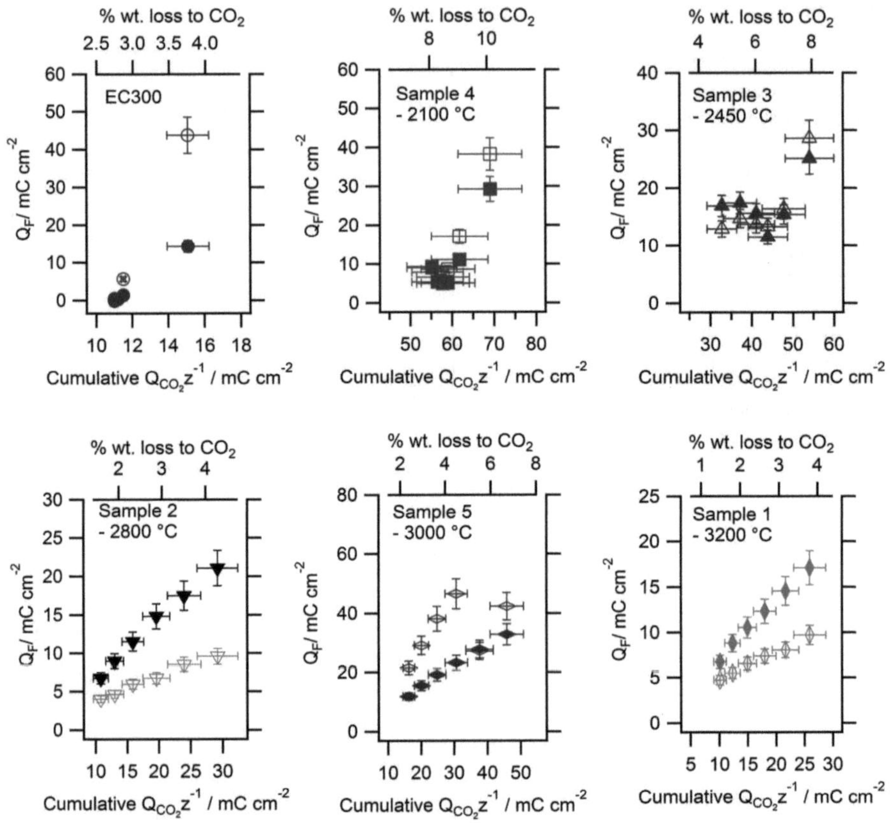

Fig. 6.15 Integrated charge of the partial (*hollow markers*) and complete (*solid markers*) electrochemical COR of the carbon black samples during the phase (3) of the electrochemical treatment. The lower potential limit was progressively decreased from 1.05 to, 0.05 V_{RHE} in 200 mV steps maintaining an upper limit of 2.05 V_{RHE} in 0.5 mol dm^{-3} Three trends are observed, either the COR is: (*1*) re-activated (EC300), (*2*) enhanced (Samples 3–4) or (*3*) unaffected (Samples 1–2 and 5), by lowering potential limit to <0.25 V_{RHE}

Although there is no commonly accepted mechanism describing the COR processes, previous work has stressed the importance of the oxygen containing surface functional groups on both the initial and long-term reactivity of carbon blacks [21, 33]. Often, the initial COR rates have been found to correlate positively with the oxygen content of the pristine carbon material (i.e. concentration of oxygen-containing surface functional groups), whereas various terms such as *phase-, protective-, stable- or surface- oxide, oxide intermediates* and *complexes*, have been used to try and describe the apparent influence of the more stable oxygen-containing surface functional groups on COR, which form during the course of an experiment (see discussion and references in Ref. [21]). It has been proposed previously that many of the 'apparently contradictory results and

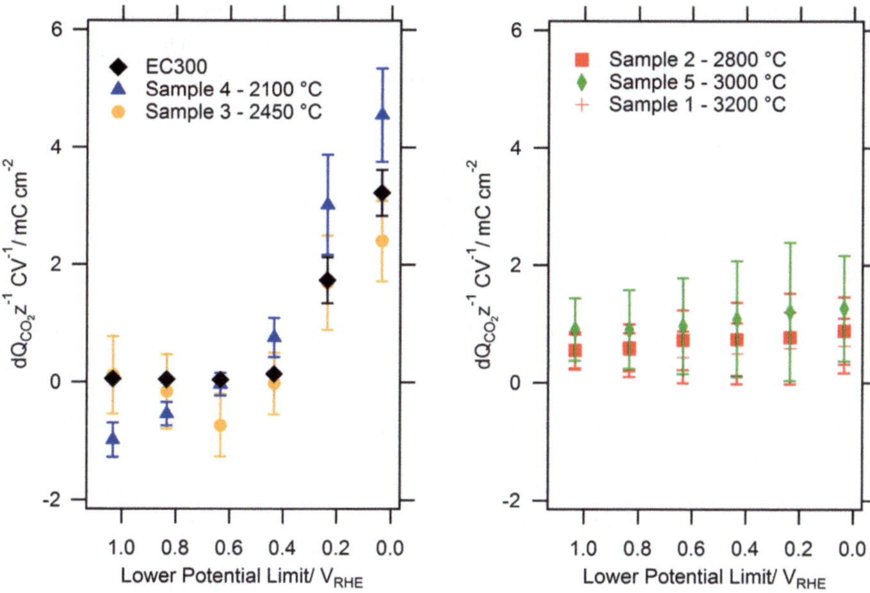

Fig. 6.16 The change in the integrated charge of the complete electrochemical COR of the HSAC carbon samples for each successive CV as a function of the lower electrode potential limit in 0.5 mol dm^{-3}. The electrochemical COR of EC300 and mildly graphitised carbon black samples are enhanced in the subsequent high potential excursion by reducing the electrode potential <0.45 V_{RHE}

interpretations (either electrochemical or gas phase oxidation [studies—prior to 1989]) could be reconciled by invoking the relative stability of these oxides as the main discriminating factor' going further to state that 'the more stable oxides might play the role of decreasing the active surface area (ASA) of the carbon' whereas 'the formation of each product molecule, CO or CO_2, is accompanied by regenerations of a free site and subsequent formation of another complex' [21] citing Ref. [55, 56]. It is this approach of considering the relative stability of the oxygen containing surface functional groups formed, and their influence on the electrochemically oxidisable surface area (EOSA) of the carbon that will be used to rationalise the trends observed in this study.

The initial rate of the COR processes on the carbon black will ultimately be determined by the nature and concentration of the electrochemically oxidisable surface functional groups that can be considered to act as nucleation sites for either the complete oxidation to CO_2, or partial oxidation to either CO or surface oxygen containing functional group(s) i.e. they define the EOSA of the carbon. If we consider that the COR to either CO_2 or CO must involve the breakage of one or more C–C bonds which will in-turn deteriorate the microstructure of the carbon generating at least one new active site, then these COR processes can be considered to either maintain or increase the EOSA (unless of course the carbon microstructure is completely consumed). Alternatively, if the partial COR proceeds to form oxygen containing surface functional group(s) that are stable (i.e. non-electrochemically oxidisable), then any further oxidation to CO_2 or CO will cease at this site, effectively decreasing the EOSA and consequently lowering the COR rate. The long-term electrochemical oxidation processes of the carbon black, therefore, will be determined by the nature of the oxygen containing surface functional groups that are formed after the initial oxidation of the pristine carbon sample, which in turn will be influenced by the carbon black microstructure, and experimental conditions.

In the first trend observed for the EC300 carbon, the complete and partial COR processes came to an abrupt end following relatively fast initial COR rates. A sceptical interpretation may have been (given the nature of the abrupt end to the COR processes) that the carbon sample is simply lost from the WE substrate; however, in phase (3) of the electrochemical treatment we observed that this was simply not the case, whilst the quantity of CO_2 evolved is far from insufficient ($\sim$4 % wt.) to account for the consumption of the sample. The EOSA of the EC300 carbon is simply diminishing to zero during the initial potential sweep. One partial explanation for this observation may be that only a certain proportion of the EC300 carbon (given the in-homogeneous nature of carbon blacks) is readily electrochemically oxidised, ceasing the COR once consumed. In previous gas phase [57] and fuel cell studies [33], for example, it has been reported that following the oxidation of the carbon black, the more amorphous core of the carbon structure appears to have preferentially oxidised over the graphene outer layers. The outer carbon layer of the aciniform structure consequently remains largely intact and a hollow structure is formed. This phenomena has been used to explain the apparent decrease in the COR of a non-graphitised carbon-black (Vulcan XC-72) in the PEMFC environment [7] and can also be considered a contributing factor to the decrease in the EOSA (and COR rate) in this study. The significant partial COR charge, however, suggest another contributing factor toward the abrupt end in the COR of the EC300 carbon. If the COR of the carbon surface favours the formation of stable non-electrochemically oxidisable oxygen-containing surface functional groups (as suggested by the partial COR charge and also phase (3) of the measurement), then the EOSA of the EC300 carbon would decrease slowing the COR. The abrupt end in the COR on the EC300 carbon would suggest that the effect is so strong that it completely diminishes the EOSA. This behaviour would be analogous to the description of a protective oxide described in a previous carbon corrosion study [58].

The COR rates of the more heavily graphitised carbons ($\geq$2,800° C) on the other hand was initially rather low, but progressively increased with cumulative electrochemical oxidation, suggesting that the EOSA of the carbon is increasing during the course of the experiment. An increase in the EOSA could conceivably arise from a number of consequences of the COR processes. For instance, the breakage of C–C bonds would deteriorate the ordered structure of the graphene planes, conceivably opening sub-surface voids (as suggested in Ref. [59]) forming micropores [60] and increasing the total surface area, along with the EOSA. Alternatively, it has also been suggested that the electrochemical oxidation of carbon black only develops after a critical concentration of oxygen-containing functional groups on the surface [21], although it is not entirely clear why this should be unless it influences the nature of the oxygen-containing functional groups that are formed. Either way, the apparent increase in the COR rate of the progressively electrochemically oxidised-heavily graphitised carbons (≥2,800° C) is analogous observations in a previous DEMS study which found that the CO_2 generated at 1.5 V_{RHE} from oxidised-graphitised carbon sample is greater than that of the pristine graphitised (as well as raw) carbon black counterpart [30]. In the PEMFC environment, however, the complete COR rates of graphitised carbon containing MEAs have been observed to be rather constant [7]. The mildly graphitised carbons (≤2,450° C) meanwhile appear to exhibit a combination of both the COR tendencies shown for the pristine EC300 and heavily graphitised carbons, which became more evident in phase (3) of the electrochemical treatment.

The COR trends observed in phase (2) of the experiment are further complicated by the potentially significant influence of the lower potential limit, or rather the reduction of the apparently stable oxygen-containing surface functional group(s)proposed to have formed during phase (1) and (2) of the experiment. The partial and complete COR of the EC300 and mildly graphitised HSAC supports may be re-activated or enhanced by lowering the electrode potential to <0.240 V_{RHE}. Although the stable oxygen-containing surface functional group(s) cannot be electrochemically oxidised, it appears that by reducing these functional groups, the carbon surface can then be completely oxidised via an alternate pathway, avoiding the formation of the stable oxygen-containing functional group. The partial and complete CORs on the heavily graphitised HSAC supports meanwhile, were in contrast unaffected by the lower potential limit, suggesting that the partial COR of the heavily graphitised carbon (with more graphitic character) does not form these non-electrochemical oxidisable oxygen-containing surface functional group(s) which upon being reduced may be oxidised in the subsequent high potential excursion. The formation of the passivating non-electrochemically oxidisable surface functional group(s) has a significant impact on the total COR charge evolved over the course of the electrochemical treatment and will be discussed later.

It is premature to begin discussing the identity of the oxidising and more stable oxygen-containing functional groups, however, if we consider that the carbon black microstructure are essentially organic molecular structures (albeit rather large ones) then an understanding of carbon corrosion mechanism(s) can conceivably utilise much from the field of organic electrochemistry in order to aid in the understanding of the relative reactivities of the surface oxygen-containing functional groups. For example, the oxidation of the carboxylic acid functional group is well-studied in organo-electrosynthesis because of its importance in the Kolbe reaction. Carboxylic acids may electrochemically oxidise to yield a CO_2 molecule and either a radical or carbocation intermediate, the latter of which may then react further via nucleophile substitution, deprotonation or C–C bond cleavage/rearrangement forming a new functional group [61]. Alcohols meanwhile can be oxidised to carboxylic acids or ketones [62] whilst hydrocarbons may undergo direct anodic oxidation under certain conditions [63]. Lactones on the other hand are not electrochemically oxidisable (or there are no examples) but can be reduced to carboxylic acids, diols or aldehydes [64]. These are just a handful of the reactions that surface functional groups present on the carbon surface can undergo, and a glance inside Ref. [65, 66] will quickly demonstrate just how complex the COR processes may be.

After establishing some understanding of the dynamic behaviours of the COR processes, along with the influence of the reduction of the carbon, it is now appropriate to evaluate the overall effect of the electrochemical treatment on the carbon samples in the following section.

6.3.3.5 Overall Effect of Electrochemical Oxidation Treatment

The total quantity of the complete and partial COR charges evolved during the electrochemical oxidation treatment are summarised as a function of the heat-treatment temperature in Fig. 6.17.

The complete COR of the graphitised HSAC supports resulted in a carbon weight loss of between 10 and 4 % wt. with generally higher resistance observed for the more heavily graphitised samples. The ratios of the complete to the partial COR charges on the graphitised samples are ~1:1 with the exception of sample 5 (3,000° C) where the partial COR current appeared to be greater. The carbon weight loss of the non-graphitised EC300 carbon black to CO_2 was overall comparable to the more heavily graphitised samples. When normalised to the BET area of the samples, the specific complete COR charge evolved from the EC300 is in fact less than half that of the most graphitised materials (≥2,800° C).

At the end of the electrochemical treatment the CV of the HSAC support samples were also evaluated, analogous to the initial measurement used to determine the apparent double layer capacitance (as described previously in Sect. 6.4.1). The capacitance determined using CV of the pristine EC300 is compared to the capacitance after the electrochemical oxidation treatment in Fig. 6.18a.

The apparent double-layer capacitance of the EC300 sample has not only clearly increased significantly but a clear peak corresponding to reversible oxidation of the carbon surface has appeared at ~0.6 V_{RHE} following the electrochemical treatment is formed. The difference between the capacitance before and after the electrochemical oxidation procedure for each of the carbon samples is given in Fig. 6.18b. Here we observe that the apparent double-layer capacitance increases for all samples, coinciding with the formation of a peak at 0.6 V_{RHE} which is a characteristic of the CV of a number of carbons and is attributed to the redox active quinone/hydroquinone surface functional groups [67]. In general, the size of the apparent double-layer capacitance and peak at 0.6 V_{RHE} decreases with increasing graphitisation temperature. Finally, the capacitance at 0.2 V_{RHE} and the integration of the peak at 0.6 V_{RHE} (background capacitance at 0.2 V_{RHE} subtracted) is shown to be directly correlated in Fig. 6.18c.

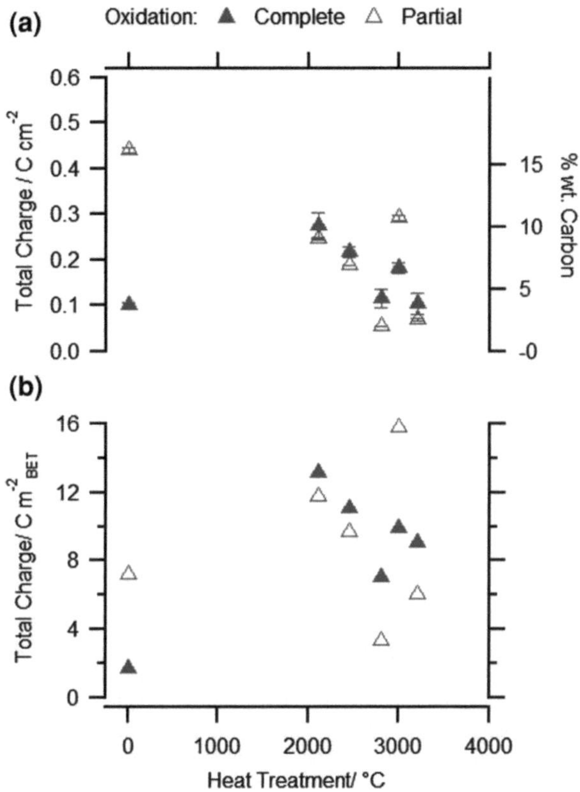

Fig. 6.17 The total complete and partial oxidation charge evolved during the electrochemical treatment versus heat treatment temperature (**a**) and corresponding data normalised to BET surface area (**b**).The complete and partial COR charges *decrease* with increasing graphitisation temperature. The partial COR dominates on pristine EC300 exhibiting less COR charge normalised to BET are than the graphitised materials (EC300 sample normalised to 85 μg_c cm^{-2})

The relationship between the total complete and partial COR charges evolved during the course of the electrochemical treatment and the quinine/hydroquinone peak charge for each of the HSAC samples are given in Fig. 6.19a [and normalised to their BET areas in (b)].

It appears that the peak charge (and in turn the increase in capacitance) is directly correlated to the partial and complete COR charge evolved during the electrochemical oxidation treatment with the exception of two points, the complete COR charge of the EC300 sample, and the partial COR of sample 5 (3,000° C). The partial COR charge measured for carbon sample 5 is believed to be an anomaly arising from changes in the WE substrate background, a contributing factor that will be discussed in the next section.

6.3.3.6 Discussion of the Overall Effect of the Electrochemical Oxidation

The overall mass normalised complete COR charge evolved during the course of the electrochemical treatment followed in the order of mildly graphitised (≤2,450° C) > heavily graphitised (≥2,800° C) ~ pristine EC300 (>BET normalised), which is in stark contrast to the complete COR charge evolved following

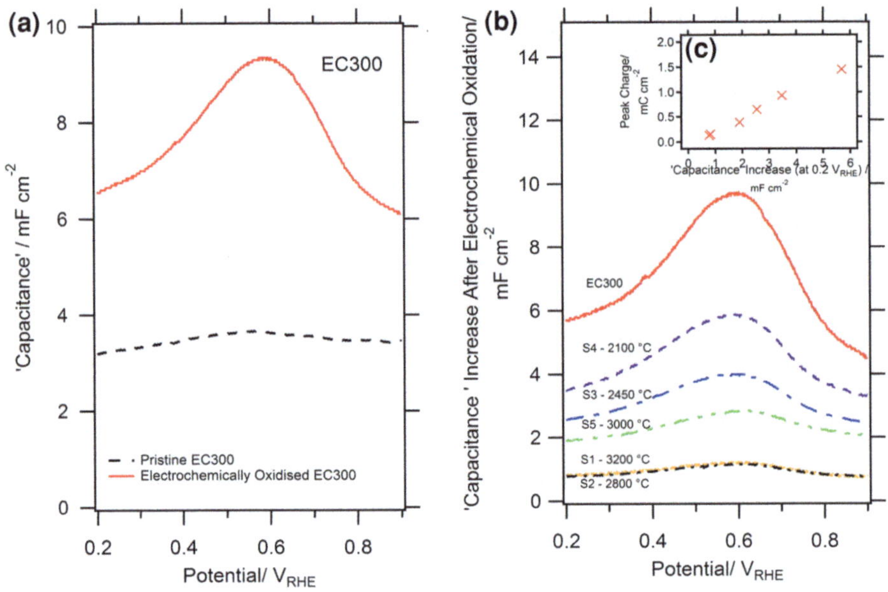

Fig. 6.18 The apparent double layer capacitance for EC300 carbon measured before and after electrochemical oxidation (**a**) and increase in apparent double-layer capacitance for each carbon black following electrochemical oxidation (**b**). The double-layer capacitance of the HSAC sample substrate is observed to have increased following the electrochemical oxidation treatment. The capacitance at 0.2 V_{RHE} scales linearly with the integrated charge of the peak observed at 0.6 V_{RHE} (**c**) attributed to the reversible oxidation of surface oxide functional groups

the initial potential sweep: pristine EC300 >mildly graphitised >heavily graphitised. Clearly, an initially fast complete COR rate (although very significant when supporting catalyst particles) does not necessarily dictate the long-term electrochemical COR behaviour of the carbon black. The lower overall total COR charge arising from the EC300 compared to the mildly graphitised (≤2,450° C) can be attributed to the rapid formation of the passivating non-electrochemically oxidisable functional group which quickly reduces the EOSA, as suggested by the significant partial COR charge that does not lead to CO_2 during the initial potential excursion. The lower complete COR charge evolved on the heavily graphitised carbons (≥2,800° C) meanwhile can be simply attributed to its superior resistance to electrochemical oxidation; no evidence to suggest that the passivating non-electrochemically oxidisable functional group(s) is formed (which when reduced <0.24 V_{RHE} can be oxidised to CO_2 via an alternative pathway). Although on the other hand the mildly graphitised carbon exhibited evidence to suggest that the non-electrochemically oxidisable species had formed in phase (1) and (2), it was not to the extent that the EOSA was completely diminished and consequently (combined with inferior resistance to oxidation) a significant amount of complete and partial COR occurred. Depending on the carbon, therefore, the non-electrochemically oxidisable oxygen-containing functional groups may have a significant

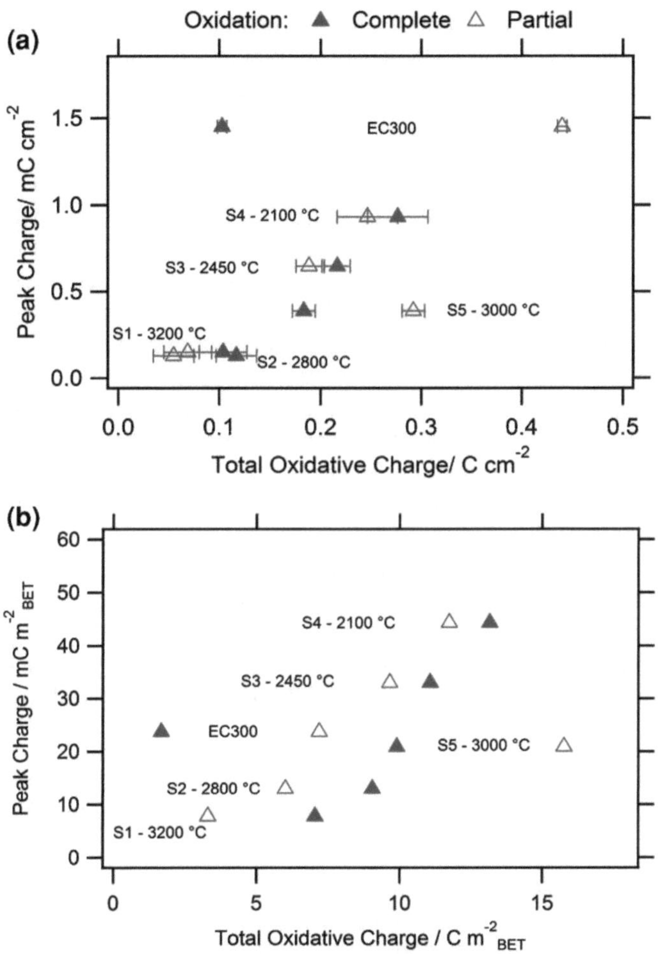

Fig. 6.19 The quinone/hydroquinone redox peak charge plotted as a function of the total partial and complete COR charge evolved during the electrochemical treatment (**a**). Corresponding data normalised to BET surface area (**b**). The peak integration scales with the total partial COR current

influence on the long term oxidation processes, however, its stability is dependent on the lower electrode potential limit.

Finally, a significant increase in the apparent double-layer electrode capacitance, along with the appearance of a peak at 0.6 V_{RHE} (attributed to the hydroquinone/quinone redox processes) appeared after the electrochemical treatment was observed for all the carbon samples. This increase, along with the appearance of a peak at 0.6 V_{RHE} is in agreement with the observations of many studies which observed as an increase in the apparent double-layer capacitance and the formation of a peak at ~0.6 V_{RHE} following electrochemical oxidation [18, 20, 22, 25, 26, 68]. In this study, a close relationship between the quinone/hydroquinone peak at

0.6 V_{RHE} (as well as the increase in the apparent double-layer capacitance at 0.25 V_{RHE}) and the total COR charges evolved during the electrochemical measurement was revealed. There were, however, two discrepancies: (1) the complete COR charge of the EC300 was too low, and (2) the total partial COR charge for Sample 5 heat treated at 3,000° C was somewhat large, and inconsistent with similarly heat-treated Samples 2 and 1 (2,800 and 3,200° C). If, however, we closely evaluate the CV and MSCV of Sample 5 (3,000° C) during phase (3) of the electrochemical treatment (see Appendix B Fig. B.2.) it appears that the OER current is much more significant than all other measurements indicating a change in the WE substrate background contributions. This highlights the limitation of the DEMS measurement, in the sense that the partial COR can only be deduced by subtracting the polycrystalline Au WE substrate current and I_{CO2} from the overall faradaic electrode current. Any unexpected change in the WE substrate background will therefore invalidate the partial COR charge deduced, which appears to be the case for this particular measurement. For example, the leakage of electrolyte around the sides of the WE substrate during the course of the measurement increases the polycrystalline Au WE substrate ECSA and henceforth the OER current, leading to the overestimation of the partial COR charge. It can therefore be concluded that the quinone/hydroquinone redox peak (and/or apparent double layer capacitance) scales proportionally with the partial COR and not the complete COR, as suggested by the EC300 sample. The final section of this study will briefly compare the observations of the pure EC300 sample, with that of EC300 HSAC supported Pt catalysts.

6.3.4 Electrochemical Oxidation of HSAC Supported Pt Catalysts

After establishing the electrochemical oxidation tendencies of the pure EC300 HSAC support in Sect. 6.4.3, the study was briefly extended to the investigation of EC300 HSAC supported Pt nanoparticle catalysts a comparison of the COR processes. An example measurement that shows the various electrochemical reactions that occur during a high potential excursion for the EC300 HSAC supported Pt 5 nm nanoparticle catalyst (analogous to phase (1) of the experiment in Sect. 6.4.3.1) elucidated using DEMS is illustrated in Fig. 6.20.

During the potential excursion to 1.8 V_{RHE} the electrode current observed <1.35 V_{RHE} is dominated by currents arising from the multilayer oxidation of the Pt surface as well as contributions from capacitive currents. As the WE potential rises above 1.35 V_{RHE} the oxidation of the HSAC support to CO_2 commences, closely followed by the evolution of O_2 arising from OER on the Pt surface commencing at 1.5 V_{RHE}. At potentials exceeding 1.5 V_{RHE} the faradaic electrode current is clearly dominated by the OER, arising from the rather fast OER rate on Pt. The significant quantity of O_2 produced during the OER on Pt imposes a constraint on the upper potential limit that can be studied in comparison to the pure carbon sample as discussed previously in Sect. 6.4.2. Consequently, the study of the HSAC supported Pt catalysts was limited to and upper potential of 1.8 V_{RHE}.

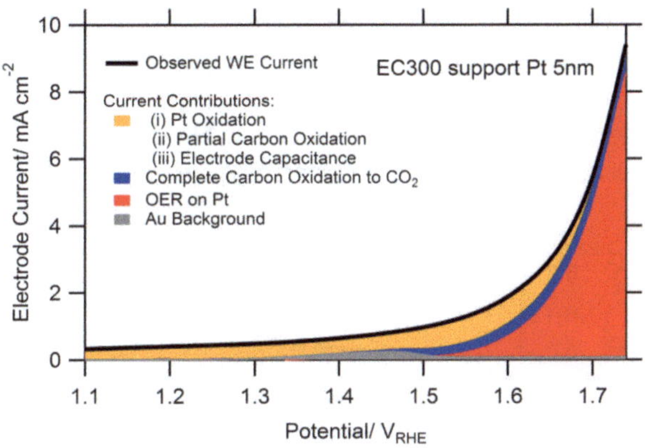

Fig. 6.20 The origins of the observed current during the initial potential excursion above 1.1 V_{RHE}. Tanaka 5 nm EC300 HSAC supported catalyst on Au substrate (50 µg cm^{-2}, 0.5 mol dm^{-3} H_2SO_4, 5 mV s^{-1}). The electrode currents above 1.4 V_{RHE} are dominated by the OER on the Pt nanoparticles with only a relatively small proportion of current arising from the complete COR to CO_2

In Fig. 6.21a the behaviour of the HSAC supported Pt 1 m catalysts to repeated high potential excursions is shown in an experiment analogous to that of phase (2) in Sect. 6.4.3.2 of the electrochemical treatment. Note that in this measurement the number of potential excursions is extended to ten rather than four in the previous electrochemical treatment (see Appendix B Figs. B.3 and B.4 for analogous measurements on HSAC supported Pt5 and 3 nm catalysts).

During the repeated LSVs of the HSAC support Pt catalyst, the faradaic electrode current is observed to decay rapidly with each subsequent potential sweep, corresponding to a significant decrease in complete COR (I_{CO2}) of the EC300 supports, along with the OER (I_{O2}) on the Pt nanoparticles. The relative decay in the charges of the reactions (including the partial COR) for each subsequent sweep are plotted as a function of the CO_2 evolved in Fig. 6.21b. In this figure we observe that the OER initially dominates the total charge evolved during the first two potential excursions but is less substantial in subsequent sweeps where the partial and complete COR charges become more significant as the total electrode charge diminishes to almost zero.

In the final phase of the electrochemical treatment of the EC300 HSAC supported Pt 1 nm catalyst, the lower potential limit was reduced to 0.05 V_{RHE} (analogous to phase (3) in Sect. 6.4.3.4) to observe the effect of reducing the HSAC support on the subsequent potential excursion to 1.8 V_{RHE}, as given in Fig. 6.22.

By lowering the potential to 0.05 V_{RHE} the complete COR of the EC300 HSAC support is re-activated as indicated in the electrode current and I_{CO2} in the subsequent high potential sweep, forming a peak at ~1.7 V_{RHE} analogous to that observed on the pure EC300 carbon in Fig. 6.12 Sect. 6.4.3.4. Furthermore, a small additional CO_2 peak is observed at 0.75 V_{RHE} whilst the OER seems to also be enhanced in the first subsequent potential sweep at 1.5 V_{RHE}, shifting to the high potential of 1.6 V_{RHE} in the following three sweeps. It should be noted, however, that the peak at 0.75 V_{RHE} is not necessarily reproducible (see Appendix B, Figs. B.3 and B.4) and that it cannot be guaranteed that no residual CO was present in the electrolyte because CO-Stripping measurements were also

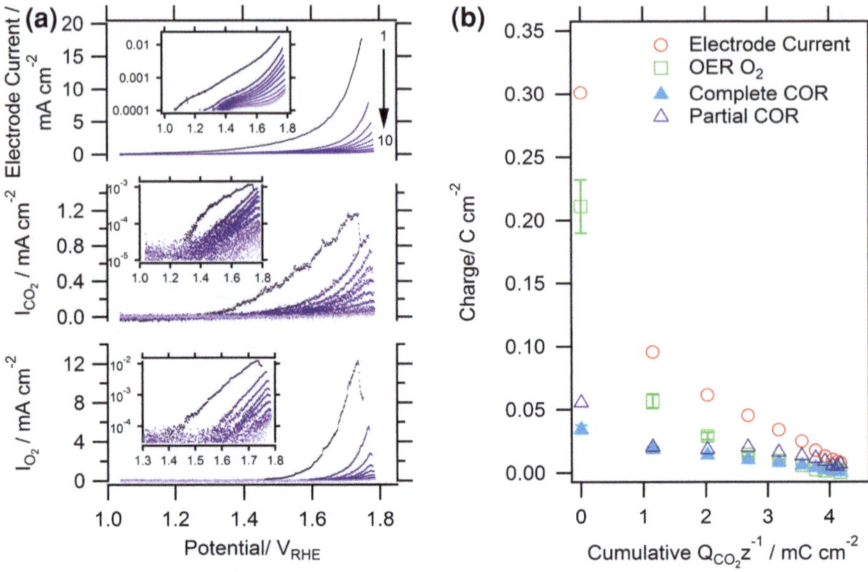

Fig. 6.21 First ten LSVs of the electrochemical COR of a (EC300) HSAC supported Pt 1 nm catalyst (50 μg_{Pt} cm^{-2}) on polycrystalline Au substrate in 0.5 mol dm^{-3} H_2SO_4 between 1.05 and 1.8 V_{RHE} (**a**). Corresponding charges of the electrochemical processes during each sweep plotted as a function of the cumulative CO_2 charge evolved (**b**). Following a rather high amount of COR in the first CV, the COR gradually ceases in subsequent CVs between 1.05 and 2.05 V_{RHE}

performed prior to this particular measurement. Of course, the COR is expected to produce some quantity of CO as well as CO_2.

6.3.4.1 Discussion of HSAC Supported Pt Catalyst Oxidation Processes

By investigating the electrochemical oxidation tendencies of the HSAC supported Pt 1 nm catalyst three important observations were made. Firstly, during the initial high potential excursion to 1.8 V_{RHE} the faradaic current contains contributions from the complete COR to CO_2 of the HSAC support, but is dominated by the OER on the Pt nanoparticles. Secondly, the electrode, complete COR and OER charges all decreased significantly during repeated excursions, decaying to almost zero after ten potential sweeps and thirdly, by lowering the electrode potential to 0.05 V_{RHE} the complete COR of the HSAC support was significantly reinitiated in the subsequent high potential sweep analogous to the pure EC300, with an additional peak feature at 0.8 V_{RHE} that may be observed.

In this study of EC300 HSAC supported catalysts no CO_2 was observed during the initial excursion until potentials >1.35 V_{RHE}, analogous to the observations

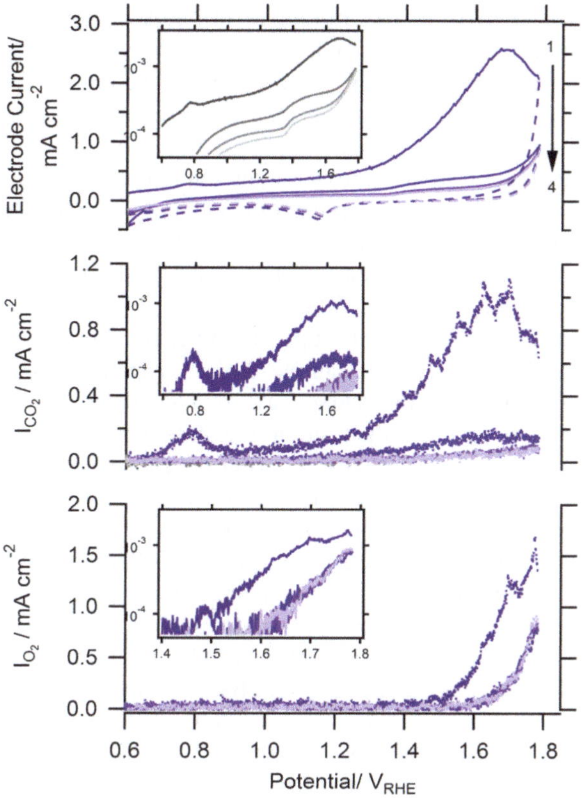

Fig. 6.22 A plot illustrating the re-activation of the complete COR of a Pt 1 nm HSAC supported catalyst [following phase (2) of the electrochemical treatment] by reducing the WE potential to 0.05 V_{RHE}, analogous to the effect observed on pure EC300 carbon (see Fig. 6.12)

on the pure EC300 carbon black in section Fig. 6.5. In previous DEMS studies, however, two additional peaks have been seen in the mass ion current for CO_2 at ~0.8 [22, 48] and 0.95 V_{RHE} [22] during the CVs of the HSAC supporting Pt catalysts that were not present for pure carbon blacks NORIT BRX [22] and Vulcan XC–72 [48]). Again, it is difficult to compare between the findings of these DEMS studies because previous they are largely qualitative, the carbon blacks studied are different (which can have a wide range of bulk and surface properties) and the detection limit of the DEMS instrument used is not given. Furthermore, the initial potential excursions were not shown in either of the previous DEMS studies i.e. the carbons were already electrochemically oxidised.

Repeated potential excursions from between 1.05 and 1.8 V_{RHE} resulted in the rapid decrease of both the faradaic, OER and complete COR current during subsequent sweeps. The decrease in the OER current can be attributed to two processes: (1) the decrease in the OER kinetics of the supported Pt nanoparticles decreases and (2) more importantly, the overall ECSA of the supported Pt decreases i.e. Pt nanoparticles detach from the HSAC support during these high potential excursions. These conclusions are based on kinetic studies of the OER on Pt which demonstrate the overpotential decreases with the extended oxidation of the Pt surface (i.e. multilayer Pt-Oxide formation) and observations using non-destructive TEM which show that Pt particles have detached from the HSAC support following extended potential cycling to a potential of 1.4 V_{RHE} [16, 17]. Another important observation in these repeated potential excursion measurements was that the complete COR to CO_2 was also observed to decay completely, analogous to the measurement on the pure EC300 which is attributed to the formation of a non-electrochemically oxidisable oxygen-containing surface functional group (see Sect. 6.4.3.5), although the processes required additional potential sweeps (ten vs. one) owing to the lower, upper electrode potential limit (1.85 V_{RHE} vs. 2.05 V_{RHE}).

The EC300 HSAC supported Pt catalyst not only exhibited the same behaviour as the pure EC300 sample during repeated high potential excursions, but also responded in the same way to a decrease in the potential to 0.05 V_{RHE} i.e. following the reduction of the passivated carbon surface, the electrochemical oxidation is re-initialised during the subsequent sweeps. On the HSAC supported Pt catalysts the significantly large peak at 1.7 V_{RHE} (observed on the pure EC300) was accompanied by a second peak at ~0.75 V_{RHE}, analogous to that observed in previous DEMS studies on NORIT BRX and Vulcan XC–72 [22, 48] carbons supporting Pt, that were not present during the initial potential excursion of the pristine EC300 support Pt catalysts in this study. The correlation of the small peak at 0.75 V_{RHE} and the peak of CO-Stripping measurement led to the interpretation that the Pt is oxidising CO_{ads}. Based on the additional CO_2 peak observed on the Pt supporting carbons, it was concluded in both previous studies that Pt catalyses the corrosion of the carbon support [22, 48]. The problem with this conclusion is that it is not supported by other experiments in electrochemical half-cells which in general require >1.2 V_{RHE} to corrode HSAC supported Pt catalysts.

At least part of this discrepancy arises from the definition of carbon corrosion. Carbon corrosion (whether induced by temperature or electrode potential) arises from the oxidation of the carbon surface, which may be via the complete COR to CO_2, or the partial COR to either surface oxygen functional groups or CO. Regardless of the corrosion product [whether CO_2, surface oxygen functional group(s), CO or even volatile organic compounds (VOCs)] the result of the carbon oxidation is essentially the breakage of C–C bonds, which deteriorates the microstructure of the carbon black reducing its ability to act as a support for finely dispersed Pt catalyst particles. Using this definition, if the role of the Pt supported is to purely catalyse the oxidation of CO species (generated by the partial COR) to CO_2, then Pt cannot be considered to catalyse the corrosion of the carbon support. The end gaseous product (whether CO_2 or CO that is later oxidised to CO_2 via Pt) is of little significance to the corrosion of the carbon black because the corrosion occurred in the partial COR step. In contrast, the subsequent oxidation of CO to CO_2 is very significant to the observations of corrosion experiments which only monitor the CO_2 produced.

6.3.4.2 Relationship between HSAC Support Oxidation and ECSA of Pt

Regardless of the COR rates on the HSAC support, the most crucial relationship in terms of PEMFC durability is the relationship between the quantity of COR and the ECSA of the finely dispersed Pt particles. In the preceding electrochemical treatment, the ECSA of the HSAC supported Pt 1 nm catalyst was determined using CO-Stripping prior to phase (1), following phase (2) and at the end of the electrochemical treatment [after phase (3)]. The relationship between the ECSA of Pt and the CO_2 evolved in the complete COR is given in Fig. 6.23. The ECSA values for the 3 and 5 nm for an analogous measurement are also given; however, phase (2) of the electrochemical treatment was limited to only four CVs and the CO-Stripping measurement was only performed after phase (3) of the electrochemical treatment. The two treatments (A and B) are therefore not directly comparable.

After phase (2) of the electrochemical treatment for the HSAC supported Pt 1 m catalysts (in Treatment A with ten sweeps to form the passivating oxide) the total ECSA of the Pt nanoparticles decreased by a significant 72 % corresponding to a carbon % wt. loss of only 4 %. After then reducing the electrode potential to 0.05 V_{RHE} to re-initialise the COR processes in the subsequent sweep (performed twice) evolving another 4 % wt. in carbon, the total ECSA of the Pt 1 nm particles is largely unaffected decreasing by only a further 13 %. The losses of ECSA for the 3 and 5 m catalysts are meanwhile ~40 and ~20 % after ~2 % wt. loss of carbon via the complete COR to CO_2.

The key objective of Fig. 6.23 is to demonstrate the ability of DEMS to correlate the quantity of CO_2 evolved during the complete COR (via the integration of the I_{CO2} during the CV) with the decrease in the total ECSA of the support Pt nanoparticles (determined using CO-Stripping). In doing so, it is observed that the ECSA loss of the supported Pt catalysts is diminished significantly, even with rather low quantities of carbon weight loss (via the complete COR) highlighting that for all the discussion of the long-term COR processes, the initial electrochemical oxidation processes of the carbon is indeed most significant; the large quantities of CO_2 evolved in phase (3) of the experiment had little impact on the ECSA.

6.3.5 Future Applications of DEMS in the Study of the COR

In order discuss the potential future of DEMS in the study of the COR, it is crucial to first establish how the limitations of the instrument impose certain restrictions on the experimental design, primarily the electrode potential region of investigation. In this study, the electrochemical oxidation of the HSAC support was explored at rather high potentials up to 2.05 V_{RHE} which is ~0.3–0.6 V higher than the potential induced at the cathode during the PEMFC start-stop events ($\leq$1.75 V_{RHE}) and a further ~0.8 V greater than those experienced when the

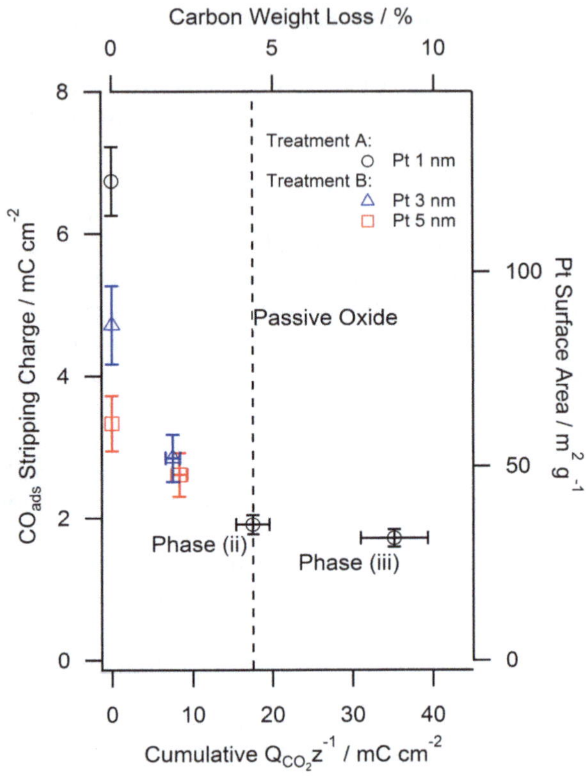

Fig. 6.23 CO Stripping charge vs. cumulative CO_2 charge evolved during the oxidatioin of various HSAC supported Pt nanoparticle catalysts. The CO-Stripping charge for the 3 and 5 nm catalysts are normalised to 14 µgPt cm^{-2} (whilst the CO_2 evolved is for 50 µg$_c$ cm^{-2} support). There are two key observations: (*1*) a carbon weight loss of only 3–4 % results in a significant ECSA loss of the supported Pt particles and (*2*) the passive oxide is formed at 4 % wt. resulting in a 72 % loss in the ECSA. The reduction and oxidation of the passive oxide does not appear to further influence the loss in ECSA

PEMFC is idle (≤1.2 V_{RHE}). Extrapolating the observations in this study down to more PEMFC relevant potentials therefore, is not necessarily straightforward. The possible region of study using DEMS is determined by two aspects of the experiment: (1) the rate of the complete COR of the carbon black sample, and (2) the I_{CO2} detection limit, equivalent to a carbon weight loss (to CO_2) of 2.8 ug$_c$ hr^{-1}or a complete COR current of ~25 µA cm^{-2} (assuming 4 electrons) for the DEMS instrument constructed in this thesis. Although DEMS was able to observe CO_2 at potentials as low as 1.35 V_{RHE} (at the detection limit) on the pristine EC300 carbon black, for the reliable quantification of CO_2 (determined by the quantification limit ~250 µA cm^{-2}) requires potentials approaching 1.6 V_{RHE}.

Although the sensitivity of the DEMS instrument imposes certain constraints on the experimental parameters that may be used, the ability to differentiate between the origin of the electrode currents that arise from the complete COR to CO_2, the OER on either the polycrystalline Au or Pt nanoparticles and by deduction the partial COR is no doubt invaluable. By exploring potential up to 2.05 V_{RHE}, this study revealed clear COR processes and trends in the behaviours of the EC300 and graphitised carbon blacks samples which exhibited interesting and unpredictable behaviour, which most likely generates more questions than answers. Hopefully, this work will provide the foundation for future DEMS and combined electrochemical half-cell studies on the COR because significantly more need be done to improve our fundamental understanding of the COR processes.

This DEMS study exclusively employed CV as the technique to analyse the electrochemical behaviour, however, having established some understanding of the COR tendencies of the carbon blacks it will be interesting in future DEMS studies to employ chronoamperometry in the study of the COR. Without prior knowledge of the electrochemical oxidation tendencies of the carbons these are difficult to plan and interpret, particularly when by measuring the COR the surface of the carbon black is changed. Quantitative studies using DEMS would likely, however, require rather high complete COR rates and henceforth high electrode potentials to precisely measure the quantity of CO_2 i.e. it will likely be difficult to reliably subtract the mass ion current background.

An important area of improvement to the study of the COR processes, particularly in light of the significant partial COR current, would be to extend the QMS measurement to the detection of CO ($^{m}/_{z} = 28$) which has been observed in an earlier qualitative DEMS study [22]. Although incorporating an additional measurement channel to the QMS may seem like a rather straightforward enhancement to the methodology, it is in fact not necessarily easy to implement. If we consider the contributions to $^{m}/_{z} = 28$ mass ion current channel, such as the large N_2 background, the contributions of CO_2 fragment ions along with the sacrifices that would be made in the signal-to-noise ratios for $^{m}/_{z} = 44$ and 32 by introducing an additional channel, some effort in optimising the QMS parameters will be required. It should, however, be noted that in the study in Ref. [22] the $^{m}/_{z} = 28$ channel was not calibrated/quantified and therefore contributions from CO_2 fragments were not subtracted.

A possible area of future interest which was touched upon in this study, is for DEMS to quantitatively relate the amount of carbon black weight loss (from the observed CO_2 and perhaps CO in the future) with the ECSA decrease of the supported Pt particles (preferably using H_{upd}) during repeated high potential excursions which could be of significant fundamental interest to PEMFC development. This could also be combined with complementary non-destructive TEM measurements to observe the changes in the catalysts structure.

A quite demanding enhancement to the DEMS experiment would be to combine a EQCM with DEMS, the useful combination of which has been previously demonstrated in the study of the electro-oxidation of CO on Pt and PtRu [69]. The usefulness of EQCM alone for the evaluation and screening tool for the corrosion behaviour of carbon supports [22]. EQCM is able to monitor the change in mass of the electrode, which for carbon blacks was observed to increase $>0.25\ V_{NHE}$ typically peaking at $\geq 1.0\ V_{NHE}$ whereby the decrease is attributed to the oxidation of the carbon black. By combining EQCM with DEMS the change in the mass of the carbon black may be monitored alongside the formation of CO_2 and CO which may bring further insights into the COR processes.

Although there is much that can be done, in order to study the COR processes at more PEMFC relevant potentials, either an increase in the DEMS instrument detection limit or an increasing the complete COR rate is required. The detection limit of the DEMS instrument may be improved by enhancing the DEMS cell collection efficiency, although if we consider that it is typically ~0.3 at reasonable electrolyte flow rates (~5 μL s^{-1}), only modest gains are possible. A more likely successful alternative would be to further optimise the QMS to $^{m}/_{z} = 44$ in order to enhance the mass ion current signal-to-noise ratio. In this study both $^{m}/_{z} = 44$ and $^{m}/_{z} = 32$ were measured simultaneously, although on the pure carbon blacks O_2 is of perhaps less interest below 1.7 V_{RHE} where the OER does not to occur on the polycrystalline Au WE substrate. It may therefore be useful to measure a single mass ion channel using the QMS, which would not suffer from the data loss that arise when using multiple mass ion channels, along with finite time required for the operational amplifier to rise and fall responds to stepped changes in the mass ion current (as the mass ion channels are switched—see Chap. 3). There are, of course, other possibilities to increase the QMS sensitivity by increasing the SEM voltage (although in this study it was at the upper limit of the linear range i.e. 1,600 V) or by being more liberal with the QMS resolution parameter so that a lower amplification range on the EP 422 unit may be used.

In terms of increasing the quantity of CO_2 generated, the most conceivable option would be to increase the loading of carbon material on the WE substrate. In the current DEMS cell, the WE substrate was a 5 mm Ø disc which, at the loading density of ca. 50 μg cm^{-2} supports 10 μg of carbon. By increasing the WE substrate size to a ca. 10 mm Ø disc supporting 50 μg of carbon material, a factor of five could be gained in the CO_2 produced via to the complete COR. An increase in the WE substrate size, however, will be difficult to implement in the current dual thin-layer flow cell design. If we consider that the flow of electrolyte is not critical in COR measurements, it may be beneficial to employ a thin-layer flow cell under static electrolyte conditions which would allow the easier implementation of a large WE electrode and higher DEMS cell collection efficiencies (see Chap. 2 for more details). It may also be easier to study the COR using chronoamperometry under static electrolyte conditions because the CO_2 background is perhaps even more stable than when using the precision syringe pump, and can therefore be more reliably subtracted. Static electrolyte may, however, introduce certain practical difficulties such as the complete purging of the DEMS cell, along with the possible formation of gas bubbles at high complete COR or OER rates forming at the WE leading to loss in WE potential control. If the combined increase in the CO_2 concentration, and with the enhancement of the QMS sensitivity can achieve a factor of ten in improved detect-ability, however, then the potential region of study for the COR of the carbon blacks can in principle be decreased by ~200 mV in potential (based on the Tafel-slopes in Sect. 6.4.3.1).

Overall, the electrochemical oxidation of HSAC supports is dynamic and rather complex, requiring a number of complementary techniques and DEMS can certainly contribute significantly toward future understanding.

6.4 Conclusions

The purpose of this study was to utilise DEMS to investigate and compare the electrochemical COR processes of carbon blacks. The approach was to compare the electrochemical oxidation current using CV up to 2.05 V_{RHE}, and elucidate the contributions arising from the partial and complete COR, and OER on either the polycrystalline Au substrate or supported Pt particles using DEMS. As part of the study, a method was developed to quantify the loading of the HSAC support on the WE substrate in situ by measuring the electrode capacitance using CV and comparing this with a pre-determined gravimetric capacitance. Furthermore, a polycrystalline Au was favoured as WE substrate rather than the more common GC substrate to avoid an ill-defined I_{CO2} background.

By quantifying the CO_2 produced during the electrochemical oxidation of a pristine EC300, and five heat-treated carbon black samples (2,100–3200 °C) using DEMS, it was deduced that the complete COR occurs in combination with another reaction process. In this study, the remaining current was attributed to the partial COR that may proceed to either CO or stable oxygen-containing surface functional groups, which contributed up to ~70 % of the current on the EC300 and in general ~50 % for the heat-treated carbon blacks during the first potential excursion to 2.05 V_{RHE}. The rate of complete COR during the initial potential excursion was found to be greatest for the EC300 carbon, and less for the heat-treated samples, with a trend for decreasing rates with increasing heat-treatment temperature. The difference between the EC300 and heat-treated Sample 4 (2,100 °C) normalised to BET area, however, was less significant. The potential at which CO_2 was first observed above the detection limit of this DEMS instrument during the initial potential excursion were 1.35 V_{RHE} for EC300, 1.68 V_{RHE} for Sample 4 (2,100 °C) and ~1.88 V_{RHE} for the remaining samples (≥2,450 °C), equivalent to a carbon rate loss of 2.754 ug_c h^{-1} evolved via the complete COR, or 25 μA cm^{-2} of faradaic current (assuming 4 e^- per CO_2). The Tafel-slopes extracted from the observed I_{CO2} meanwhile, were in the range of 200 ± 30 mV Dec^{-1} above 1.8 V_{RHE}, although shallower slopes of 400 and 300 mV Dec^{-1} were observed on the EC300 and mildly graphitised HSAC in the potential region below 1.8 V_{RHE}, corresponding to electron transfer coefficient (α*n) values of between 0.143 and 0.354. In order to fully rationalise the differences in the initial COR rates, however, greater characterisation of the bulk and surface properties of the pristine carbon blacks is required.

The behaviour of the COR tendencies of the carbon samples during repeated potential excursions between 1.05 and 2.05 V_{RHE} [phase (2)] and repeated potential excursions with decreasing electrode potential limit (down to 0.05 V_{RHE}) were revealed to be rather dynamic exhibiting three trends. In phase (2) the COR processes either: (1) stopped abruptly (EC300), (2) progressively increased and then decreased (2,100 and 2,450 °C), or (3) progressively increased for each subsequent CV (≥2,800 °C). Three different trends were also observed in phase (3) of the electrochemical oxidation when the lower electrode potential limit was decreased to ≤0.24 V_{RHE}, either the COR processes were: (1) reinitiated (EC300), (2) enhanced (Samples ≤2,450 °C), or (3) unaffected, in the subsequent high potential excursions (≥2,800 °C). Although there is no established COR mechanism to explain all the experimental observations, these three trends were rationalised by considering the relative stability of the oxygen-containing surface functional groups arising from the partial COR and their influence on the EOSA of the carbon. Although not the only determining factor, the stable non-electrochemically oxidisable functional group(s) may be considered to decrease the EOSA, whereas the formation of CO_2 and CO regenerates at least one new free site, maintaining or increasing the EOSA (unless of course, the carbon is completely consumed). By reducing the non-electrochemically oxidising oxygen containing functional groups, however, the complete COR can be reinitiated (EC300) or enhanced (≤2,450 °C) during the subsequent oxidation (i.e. the EOSA is increased by reducing the carbon). The study of the COR processes of carbon blacks is overall complex, and given the difficulties in characterising the pristine carbon black materials ex-situ (let alone in situ) to try to fundamentally understand the influence of electrochemical oxidation, it is only possible to speculate at this stage on what is happening to the structure and surface of the carbon materials.

The study was then briefly extended to EC300 HSAC supported Pt catalysts, which essentially exhibited the same trends observed on the pure EC300, although a small additional peak was observed at 0.8 V_{RHE} attributed to the oxidation of CO by Pt; however, it was not necessarily concluded that Pt catalyses the corrosion of the carbon support (in contrast to previous DEMS studies) because CO can already be considered a corrosion product from the partial COR (although this is not confirmed in this study). Finally, the potential application of DEMS towards correlating the HSAC support weight loss with the loss in ECSA of supported Pt particles was briefly demonstrated, which could have particular relevance to PEMFC development.

Overall, this study revealed clear and surprising trends in the COR tendencies of the HSAC support materials, demonstrating that DEMS is very useful tool in the elucidation of the electrochemical COR processes of carbon blacks.

References

1. Zhang S et al (2009) A review of platinum-based catalyst layer degradation in proton exchange membrane fuel cells. J Power Sources 194(2):588–600
2. Haas HR, Davis MT (2009) Electrode and catalyst durability requirements in automotive pem applications: technology status of a recent mea design and next generation challenges. ECS Trans 25(1):1623–1631
3. Meyers JP, Darling RM (2006) Model of carbon corrosion in PEM fuel cells. J Electrochem Soc 153(8):A1432–A1442
4. Reiser CA et al (2005) A reverse-current decay mechanism for fuel cells. Electrochem Solid-State Lett 8(6):A273–A276
5. Kim J, Lee J, Tak Y (2009) Relationship between carbon corrosion and positive electrode potential in a proton-exchange membrane fuel cell during start/stop operation. J Power Sources 192(2):674–678

6. Tang H et al (2006) PEM fuel cell cathode carbon corrosion due to the formation of air/fuel boundary at the anode. J Power Sources 158(2):1306–1312
7. Yu PT et al (2006) The impact of carbon stability on PEM fuel cell startup and shutdown voltage degradation. ECS Trans 3(1):797–809
8. Baumgartner WR et al (2008) Polarization study of a PEMFC with four reference electrodes at hydrogen starvation conditions. J Power Sources 182(2):413–421
9. Borup R et al (2007) Scientific aspects of polymer electrolyte fuel cell durability and degradation. Chem Rev 107(10):3904–3951
10. Schmittinger W, Vahidi A (2008) A review of the main parameters influencing long-term performance and durability of PEM fuel cells. J Power Sources 180(1):1–14
11. Yoda T, Uchida H, Watanabe M (2007) Effects of operating potential and temperature on degradation of electrocatalyst layer for PEFCs. Electrochim Acta 52(19):5997–6005
12. Gasteiger HA et al (2008) Catalyst degradation mechanisms in pem and direct methanol fuel cells. In: Kakaç S, Pramuanjaroenkij A, Vasiliev L (eds) Mini-micro fuel cells, Springer, Netherlands pp 225–233
13. Patterson TW, Darling RM (2006) Damage to the cathode catalyst of a PEM fuel cell caused by localized fuel starvation. Electrochem Solid-State Lett 9(4):A183–A185
14. Chen S et al (2010) Platinum-alloy cathode catalyst degradation in proton exchange membrane fuel cells: nanometer-scale compositional and morphological changes. J Electrochem Soc 157(1):A82–A97
15. Zhang J et al (2007) Effect of hydrogen and oxygen partial pressure on pt precipitation within the membrane of PEMFCs. J Electrochem Soc 154(10):B1006–B1011
16. Maass S et al (2008) Carbon support oxidation in PEM fuel cell cathodes. J Power Sources 176(2):444–451
17. Cai M et al (2006) Investigation of thermal and electrochemical degradation of fuel cell catalysts, vol 160. Elsevier, Amsterdam, p 10
18. Nose M et al (2009) Electrochemical oxidation of highly oriented pyrolytic graphite in sulphuric acid solution under potential pulse condition. Fuel Cells 9(3):284–290
19. Kinumoto T et al (2006) Stability of Pt-catalyzed highly oriented pyrolytic graphite against hydrogen peroxide in acid solution. J Electrochem Soc 153(1):A58–A63
20. Ball SC et al (2007) An investigation into factors affecting the stability of carbons and carbon supported platinum and platinum/cobalt alloy catalysts during 1.2 V potentiostatic hold regimes at a range of temperatures. J Power Sources 171(1):18-25
21. Antonucci PL et al (1989) A comparative analysis of structural and surface effects in the electrochemical corrosion of carbons. Mater Chem Phys 21(5):495–506
22. Hung C-C et al (2011) Corrosion of carbon support for PEM fuel cells by electrochemical quartz crystal microbalance. J Power Sources 196(1):140–146
23. Mayrhofer KJJ et al (2008) Fuel cell catalyst degradation on the nanoscale. Electrochem Commun 10(8):1144–1147
24. Mayrhofer KJJ et al (2008) Non-destructive transmission electron microscopy study of catalyst degradation under electrochemical treatment. J Power Sources 185(2):734–739
25. Choo HS et al (2008) Electrochemical oxidation of highly oriented pyrolytic graphite during potential cycling in sulfuric acid solution. J Power Sources 185(2):740–746
26. Avasarala B, Moore R, Haldar P (2010) Surface oxidation of carbon supports due to potential cycling under PEM fuel cell conditions. Electrochim Acta 55(16):4765–4771
27. Choo H-S et al (2007) Mechanism for electrochemical oxidation of highly oriented pyrolytic graphite in sulfuric acid solution. J Electrochem Soc 154(10):B1017–B1023
28. Kodera F et al (2007) Electrochemical corrosion of platinum electrode in concentrated sulfuric acid. J Power Sources 172(2):698–703
29. Willsau J, Heitbaum J (1984) The influence of pt-activation on the corrosion of carbon in gas-diffusion electrodes—a dems study. J Electroanal Chem 161(1):93–101
30. Colmenares LC et al (2009) Model study on the stability of carbon support materials under polymer electrolyte fuel cell cathode operation conditions. J Power Sources 190(1):14–24

31. Ming-fang L et al (2010) Oxidation of carbon supports at fuel cell cathodes: differential electrochemical mass spectrometric study. Chin J Chem Phys 23(4):442
32. Kinoshita K (1987) Chapter 2 Section 2.1 Crystallographic structure. In: Carbon: electrochemical and physiochemical properties, Wiley, New York, pp 20–85
33. Kinoshita K (1987) Chapter 6 Section 6.2 Electrochemical oxidation of carbon blacks and graphites. In: Carbon: electrochemical and physiochemical properties, Wiley, New York, pp 316–351
34. Yu PT et al (2009) Carbon-support requirements for highly durable fuel cell operation. In: Büchi FN (ed) Polymer electrolyte fuel cell durability, Springer, New York
35. Kinoshita K (1987) Carbon: electrochemical and physiochemical properties. Wiley, New York
36. Kinoshita K (1987) Chapter 3 Section 3.1.1 Carbon-oxygen complexes. In: Carbon: electrochemical and physiochemical properties, Wiley, New York, pp 87–88
37. Kinoshita K (1987) Table 3.5 Effect of heat treatment on the distribution of surface groups on carbon blacks. In: Carbon: electrochemical and physiochemical properties, Wiley, New York, p 99
38. Wiberg GKH, Mayrhofer KJJ, Arenz M (2010) Investigation of the oxygen reduction activity on silver—a rotating disc electrode study. Fuel Cells 10(4):575–581
39. Wiberg GK, Mayrhofer KJ, Arenz M (2009) Investigation of the oxygen reduction activity of non-platinum catalysts: a RDE methodology. ECS Meeting Abstracts 902(6):349
40. Trasatti S, Petrii OA (1992) Real surface area measurements in electrochemistry. J Electroanal Chem 327(1–2):353–376
41. Kinoshita K (1987) Chapter 6 Section 6.1.1 Double-layer capacitance. In: Carbon: electrochemical and physiochemical properties, Wiley, New York, pp 293–302
42. Kinoshita K (1987) Table 6.1 Double-layer capacitance of carbon and graphite. In: Carbon: electrochemical and physiochemical properties, Wiley, New York, pp 293–302
43. Barbieri O et al (2005) Capacitance limits of high surface area activated carbons for double layer capacitors. Carbon 43(6):1303–1310
44. Taylor RJ, Humffray AA (1973) Electrochemical studies on glassy carbon electrodes: I. electron transfer kinetics. J Electroanal Chem Interfacial Electrochem 42(3):347–354
45. Kinoshita K (1987) Chapter 6 Section 6.4.1 Oxygen redcution/evolution. In: Carbon: electrochemical and physiochemical properties, Wiley, New York p 369
46. Kinoshita K (1987) Fig. 2.20 Affect of heat-treatment (2 h) on crystallite size and 002 d-spacing of vulcan XC-72. In: Carbon: electrochemical and physiochemical properties, Wiley, New York, pp 20–85
47. Kinoshita K (1987) Chapter 3 Section 3.2.2 Spectroscopic analysis. In: Carbon : electrochemical and physiochemical properties, Wiley, New York, pp 105–132
48. Kinoshita K (1987) Fig. 6.15: Effect of heat-treatment temperature on the electrochemical oxidation rate of various carbons after 1,000 min at 1.0 V in H_3PO_4. In: Carbon: electrochemical and physiochemical properties, Wiley, New York
49. Kinoshita K (1987) Fig. 6.13: Variation in QCO_2/Q_t with time for the electrochemical oxidation of carbon blacks at 1.2 V in 96 % H_3PO_4, 135 °C. In: Carbon: electrochemical and physiochemical properties, Wiley, New York
50. van Veen JAR, van Baar JF, Kroese KJ (1981) Effect of heat treatment on the performance of carbon-supported transition-metal chelates in the electrochemical reduction of oxygen. J Chem Soc Faraday Trans1: Phys Chem in Condens Phases 77(11):2827–2843
51. Kinoshita K (1987) Table 2.6: Morphology parameters and physicsal proprty of carbon blacks. In: Carbon: electrochemical and physiochemical properties, Wiley, New York
52. Li W, Lane AM (2009) Investigation of Pt catalytic effects on carbon support corrosion of the cathode catalyst in PEM fuel cells using DEMS spectra. Electrochem Commun 11(6):1187–1190
53. Kinoshita K (1987) Fig. 6.17: Tafel plots for the electrochemical oxidation of Vulcan XC-72 and Vulcan-2700 in 96 % H_3PO_4 at 100 min after start of experiment. In: Carbon: electrochemical and physiochemical properties, Wiley, New York

54. Kinoshita K (1987) Fig. 6.16: Effect of heat-treatment temperature and the 002 d-spacing on the Tafel slopes of the electrochemcial oxidation of Volcan XC-72 in 102 % H_3PO_4 at 180 °C. In: Carbon: electrochemical and physiochemical properties, Wiley, New York
55. Laine NR, FVastola FJ, Walker Jr PL (1963) J Phys Chem 67:2030
56. Ahmed S, Back MH (1985) Carbon 23(5)
57. Kinoshita K (1987) Fig. 4.4: Schematic representation of the microstructural changes to carbon-black particles during oxidation. In: Carbon: electrochemical and physiochemical properties, Wiley, New York, p 185
58. Janssen LJJ, Hoogland JG (1970) Mechanism of bromine evolution at a graphite electrode. Electrochim Acta 15(10):1677–1683
59. Pittman CU et al (1997) Chemical modification of carbon fiber surfaces by nitric acid oxidation followed by reaction with tetraethylenepentamine. Carbon 35(3):317–331
60. Stonehart P (1984) Carbon substrates for phosphoric acid fuel cell cathodes. Carbon 22(4–5):423–431
61. Torii S, Tanaka H (2001) Chapter 14: Carboxylic acids, In: Lund H (ed) Organic electrochemistry, 4th edn. pp 499–544
62. Grimshaw J (2000) Chapter 8: Oxidation of alcohols, amines and carboxylic acids. In: Electrochemical reactions and mechanisms in organic chemistry, Elsevier, Amsterdam
63. Hammerich O (2001) Chapter 13: Anodic oxidation of hydrocarbons. In: Lund H (ed) Organic electrochemistry, 4th edn. pp 471–498
64. Wagenknecht JH, Eberson L, Utley JHP (2001) Chapter 12: Carboxylic acids and derivatives. In: Lund H (ed) Organic electrochemistry, 4th edn. pp 453–470
65. Lund H, Hammerich O (2001) Organic electrochemistry, 4th edn. Organic electrochemistry, Marcel Dekker, Inc, New York
66. Grimshaw J (2000) Electrochemical reactions and mechanisms in organic chemistry. Elsevier, Amsterdam
67. Kinoshita K (1987) Chapter 6 Section 6.1.3 Potentiodynamic invetigations of surface groups. In: Carbon: electrochemical and physiochemical properties, Wiley, New York, pp 293–302. 209
68. Shao Y et al (2006) Comparative investigation of the resistance to electrochemical oxidation of carbon black and carbon nanotubes in aqueous sulfuric acid solution. Electrochim Acta 51(26): 5853–5857
69. Jusys Z, Massong H, Baltruschat H (1999) A new approach for simultaneous DEMS and EQCM: electro-oxidation of adsorbed CO on Pt and Pt-Ru. J Electrochem Soc 146(3): 1093–1098

Chapter 7
Summary

The aim of this thesis was to design, construct and demonstrate the research application(s) of a DEMS instrument. The approach to this problem began with the analysis of previous DEMS design solutions before embarking on the design and construction of the DEMS instrument created in this thesis. In doing so, the intended research application of the DEMS instrument was always kept in mind: that is, to perform fundamental research studies that approach problems which are relevant to the immediate development concerns of PEMFCs.

The setup of the DEMS instrument first required the design and construction of three crucial parts: the electrochemical half-cell, membrane interface and the vacuum system of a commercial QMS. The electrochemical half-cell constructed in this thesis was of similar design to the dual-cyclone thin-layer flow cell reported previously, whereby the WE is separated from the membrane interface, which are placed in two separate thin-layer compartments. The cell construction produced in this work, however, is considered to be rather modular and intuitive to the cell development process. Similarly to all DEMS instruments, the electrochemical cell was interfaced with the vacuum of the QMS via microporous PTFE membrane which possesses a geometric area of $0.95\ cm^2$ and a total gas flux of $0.12\ mbar\ l\ s^{-1}\ cm^{-2}$ when exposed to the DEMS vacuum chamber. The DEMS vacuum system constructed here was of a 3-stage differentially pumped design, in contrast to the two-stage design more typically employed in DEMS. The sensitivity of the QMS was enhanced using a tubular aperture to direct the gas flux from the 1st pumping stage at 10^{-2} mbar, perpendicularly into the CB-ion source of the QMS in the 2nd pumping stage. A second aperture, and addition vacuum pumping at the back of the QMS allowed a pressure of $\sim 2 \times 10^{-5}$ mbar in the CB-ion source chamber whilst maintaining the necessary operating pressure of the SEM ($<5 \times 10^{-6}$ mbar). Once a working DEMS instrument had been established, the next step was to assemble and develop the instrument control hardware and software necessary to improve the instrument functionality and usability. This was primarily achieved using a combination a NI DAQ cards, and custom programmed

S. J. Ashton, *Design, Construction and Research Application of a Differential Electrochemical Mass Spectrometer (DEMS)*, Springer Theses,
DOI: 10.1007/978-3-642-30550-4_7,

Labview software. By incorporating the required DEMS functionality into an already powerful existing custom Labview software developed in our laboratory for RDE and MWE techniques, the resulting software not only allowed meticulous control of all experimental parameters, but also performs the acquisition, synchronisation, display and logging of both the electrochemical and QMS mass ion currents in real-time. Furthermore, additional software were also developed to allow easy calibration of the QMS, and to perform common DEMS data treatment and analysis tasks.

Before performing experimentation, it was crucial to understand and overcome some of the more practical aspects of the DEMS instrument, in order to optimise both the instrument and experimental parameters, understand its limitations and develop calibration methods. For example, the large cell resistances inherent to thin-layer electrochemical flow cells combined with a small solution resistance between the WE and effective Luggin-capillary resulted in potentiostat oscillation in this DEMS cell setup. These oscillations were overcome by adding a stabilising resistor to the WE connection which artificially increased the effective solution resistance, stabilising the potentiostat. *IR*-compensation of the stabilising resistor was then achieved using a NID allowing an acceptable uncompensated solution resistance of $\sim 10\ \Omega$ with stable WE potential control. Furthermore, the performance of the DEMS electrochemical flow cell was found to be highly dependent on the electrolyte flow rate which influenced the instrument sensitivity, QMS response time and the correlation between electrochemical processes at the WE and the relevant mass -ion currents observed. Optimum flow rates were found to be between 4 and 10 $\mu l\ s^{-1}$, with potential scan rates of $\leq 20\ mV\ s^{-1}$.

The research capabilities of the DEMS instrument were first demonstrated by revisiting the MOR on a HSAC Pt and PtRu catalyst using CV, chronoamperometry and for the first time in combination with DEMS: three-dimensional voltammetry. An initial finding concerning ECSA determination was that contrary to previous reports, the CO-Stripping charge of the PtRu catalyst was not overestimated if the background CV in Argon saturated solution was subtracted prior to the peak integration. On the pure HSAC supported Pt catalyst, an extremely close correlation between the faradaic MOR current, I_{CO_2}, the conversion of methanol to CO_2 and the ability of Pt to oxidise CO_{ads} was observed using all electroanalytical techniques, in agreement with the well-established understanding that the oxidation of CO_{ads} is the RDS of the MOR. This close relationship was, however, not mirrored by the MOR on the HSAC supported PtRu catalyst. Instead, the conversion of methanol to CO_2 was found to be inversely correlated to the faradaic MOR current, which was not as closely dependent on the ability of the PtRu catalyst to oxidise CO_{ads}. Consequently, the RDS for the MOR on PtRu was concluded to be the initial dissociative adsorption of methanol on Pt, although the poisoning CO_{ads} species can still be attributed to the significant electrode potential $>0.32\ V_{RHE}$ required for the MOR to proceed under these experimental conditions. The contrasting potential dependent conversion for the MOR and the Pt and PtRu catalysts was rationalised by considering that the role of Ru is to not only adsorb OH_{ads} and facilitate the oxidation of CO_{ads} on Pt, but to also oxidise desorbed partial MOR products that are produced via the incomplete MOR on the Pt portion of the surface (Ru itself is understood not to adsorb methanol at ambient temperature ca. <40 °C). Overall, the PtRu catalyst possessed a significantly higher mass activity of 72.6 compared to 1.74 mA mg_{metal}^{-1} for Pt at 0.55 V_{RHE} using chronoamperometry, however, the comparison of these activities to other electrochemical half-cell MOR studies is not entirely meaningful because there is no commonly employed methodology, a situation that is not intuitive to DMFC electrocatalyst development, particularly where claims of novel materials with 'high' MOR activity are made.

In a second study, the application of DEMS toward the fundamental study of the electrochemical oxidation of HSAC supports (a.k.a. carbon blacks) was demonstrated by comparing the electrochemical oxidation tendencies of a pristine EC300, and five graphitised carbon black samples (heat-treated at various temperatures between 2100 and 3200 °C). As part of the study, a method was developed to quantify the loading of the carbon black on the WE substrate in situ by measuring the sample capacitance and comparing this with a pre-determined gravimetric capacitance. Furthermore, a polycrystalline Au was employed as WE substrate rather than the more common GC which was found to electrochemically oxidise to CO_2 at potentials >1.4 V_{RHE}. By quantifying the CO_2 produced during the high potential excursion using DEMS, it was observed that the faradaic current contains additional current contributions attributed to the partial COR of the carbon surface to either CO or oxygen containing surface functional groups. The electrochemical oxidation tendencies of the carbon black were compared in three experimental phases. In the initial potential excursion to 2.05 V_{RHE} (phase (i)), the COR rates were observed to follow in the order EC300 > mildly graphitised (2100 > 2450 °C) > heavily graphitised carbons (≥2800 °C). During repeated potential excursions between 1.05 and 2.05 V_{RHE} (in phase (ii)), and repeated potential excursions with a progressively decreasing electrode potential limit down to 0.05 V_{RHE} (in phase (iii)), the COR processes were revealed to be rather dynamic exhibiting three trends. In phase (ii) the COR processes either: (i) stopped abruptly (EC300), (ii) progressively increased and then decreased (2100 and 2450 °C), or (iii) progressively increased for each subsequent CV (≥2800 °C), whilst in phase (iii) the COR processes were either: (i) reinitiated (EC300), (ii) enhanced (Samples ≤2450 °C), or (iii) unaffected (≥2800 °C) in subsequent high potential excursions after decreasing the lower electrode potential limit to ≤0.24 V_{RHE}. These trends were rationalised by considering the relative stability of the oxygen containing surface functional groups and the consequence of the partial and complete COR on the EOSA of the carbon black. The overall COR charge evolved for each carbon followed in the order: mildly graphitised > heavily graphitised > EC300, in stark contrast to the initial COR rates. The study was then briefly extended to the electrochemical oxidation of EC300 HSAC supported Pt catalysts, which was dominated by the OER on the supported Pt particles. Nonetheless, the EC300 support exhibited the same trend as the pure EC300 carbon, although an additional feature at 0.8 V_{RHE} was observed during phase (iii) of the experiment which was attributed to the oxidation of CO by the supported Pt particles. Finally, the potential application of DEMS towards correlating the HSAC support weight loss, with the loss of ECSA of the supported Pt particles was briefly demonstrated, and could be of particular interest for industrial electrocatalyst development.

Overall, despite that the MOR and electrochemical COR of carbon blacks have long since been studied using a variety of experimental techniques, the DEMS instrument constructed in this thesis could, and can continue to bring new insights into both reaction processes.

Appendix A

Figures A.1 and A.2.

A.1 Approach to Determining MOR Activities

Not all guidelines given in ref. [1] for the ORR activity determination (some of which are specific to the RDE technique) are entirely applicable to the MOR, however, those that are more general are given as follows:

1. The number of active sites (N_{ECAS}) must be measured
2. Capacitive current must be subtracted in the evaluation of activity
3. Compensation schemes for the solution resistance must be used
4. The observed reaction rate and the determined N_{ECAS} must depend linearly on the applied loading
5. Normalised activities, i.e. mass activity, specific activity and/or turn over frequency (TOF) should be evaluated in a fuel cell relevant potential window vs. RHE

In principle, applying these guidelines to the determination of the MOR activities should facilitate more meaningful and fuel cell relevant comparison between prospective catalyst materials. There are, however, a number of complications present for the MOR which are less clear cut than in the ORR, particularly concerning guidelines (i), (iv) and (v).

Although the total surface area of Pt and PtRu based catalyst can be rather straightforwardly measured using CO-Stripping, in the case of PtRu this does not necessarily correspond to the number of active sites (i), which is furthermore dependent on the temperature of the reaction. At 25 °C the Ru component of the surface is not active to the adsorption of methanol and therefore only the Pt surface area possesses active sites. This is, of course, further complicated if the Ru

S. J. Ashton, *Design, Construction and Research Application of a Differential Electrochemical Mass Spectrometer (DEMS)*, Springer Theses,
DOI: 10.1007/978-3-642-30550-4,

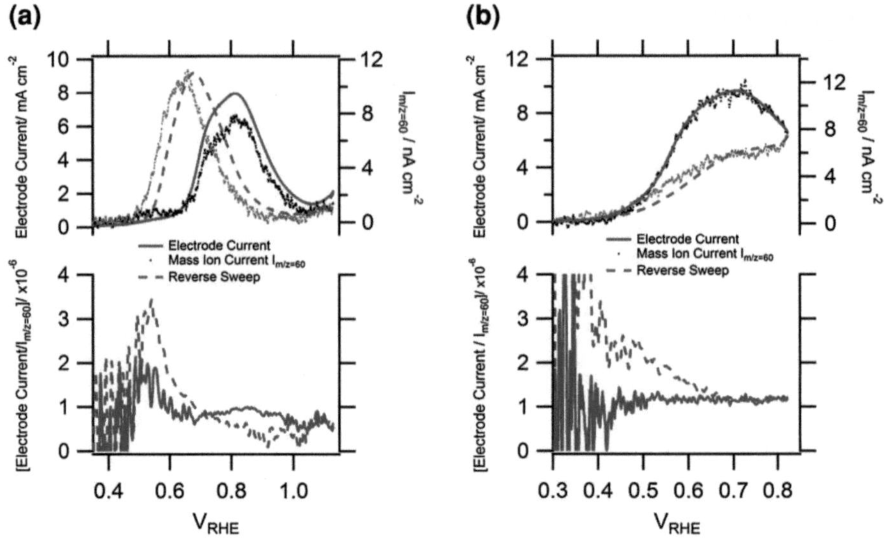

Fig. A.1 CV of the MOR on Pt (**a**) and PtRu (**b**) catalyst indicating the methylformate produced, and the ratio with the electrode current

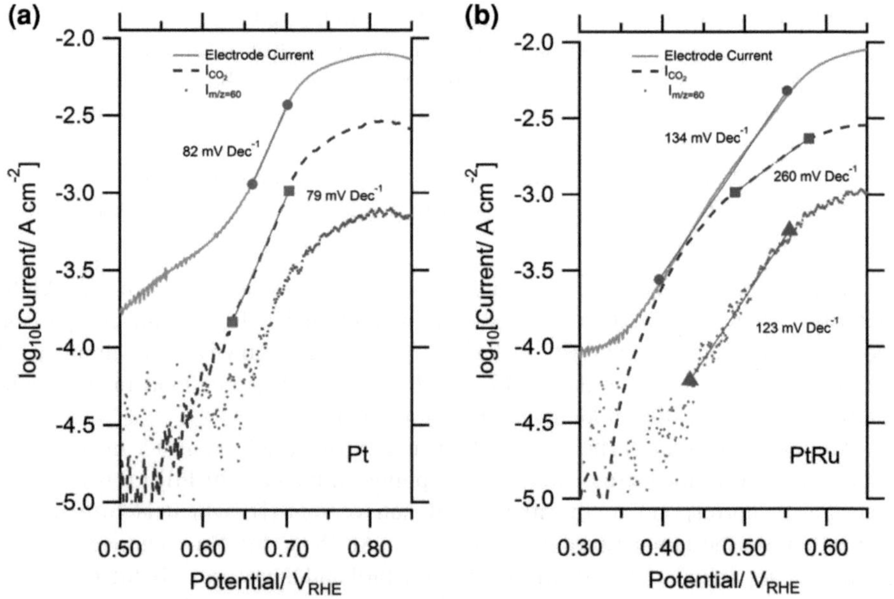

Fig. A.2 Tafel plot of the MOR on the Pt (**a**) and PtRu (**b**) catalysts. The Tafel-slope for methylformate does not correspond for the Pt catalysts, but does for the PtRu catalysts in contrast to I_{CO2}

component then contributes to the oxidation of desorbed partial MOR products. At 60 °C the situation changes and the Ru possesses some activity towards the MOR, so in such cases normalisation to the CO-Stripping charge is then of greater importance. The intended application could therefore impose different requirements on the normalisation of the catalyst material.

The number of active sites then leads us directly to guideline (v), the normalisation of activity to either mass, specific activity or TOF at a fuel cell relevant potential with reference to the RHE. The next issue is, of course, how to normalise the observed electrode current. The usefulness of the mass activity is clearly the most technologically important, particular in terms of normalising to the overwhelmingly most expensive component of a catalyst (i.e. Pt). The normalisation to the surface area determined via CO-Stripping may not be entirely useful to the development of the PtRu electrocatalyst if we consider that at ambient temperatures the Ru component is not active to the adsorption of methanol and cannot be considered an active site (at least for methanol). This is, however, perhaps less important for measurements performed at 60 °C where the Ru component possesses some activity within the potential region of interest. Unfortunately, TOF cannot be used because in general complete conversion of the MOR to CO_2 is not achieved. The definition of the fuel cell relevant potential window is also an added complication, which depends upon the operating conditions of the DMFC. For instance, the anode catalyst of a DMFC operating at >60 °C will be unlikely to exceed 0.5 V_{RHE}, however, the anode of a portable DMFC operating at ambient temperature (the most viable commercial application [2]) would operate at potentials more in line with those observed in the Tafel region of the MOR in the electrochemical half-cell measurements (i.e. <0.6 V_{RHE}). The definition of fuel cell relevant potential region is therefore dependent on the intended DMFC operating conditions. The benefits of referencing electrode potentials to the RHE meanwhile should be clear enough, but as we observed in Table 5.3 potentials are often referenced to the SCE or Ag/AgCl reference electrodes. Moreover, there is also the added problem that there is no good model MOR catalyst to which activities of prospective materials can be benchmarked too; polycrystalline Pt (which is useful for the ORR) is hardly a good benchmark for PtRu catalysts, possessing very little MOR activity <0.6 V_{RHE} at ambient temperature.

Finally, the choice of conditions and electroanalytical technique also warrants some discussion: some of my own opinions on the appropriate approach are presented in Table A.1.

The purpose of this discussion on the activity of the electrocatalysts was not to try to explain how to measure but was to rather highlight the problem, with the intention of opening the discussion and to offer some of my own thoughts on the matter; however, the development and agreement of a methodology will require discussion between different experts in the field. A documented aid to determine the activities would then be beneficial, clarifying the various techniques that can be employed and experimental considerations: such an informative guide could not make the situation any worse than it currently is.

Table A.1 List of experimental variables that must be considered when performing MOR activity measurements in an electrochemical half-cell using an RDE in acid electrolyte

Variable	Advantages	Limitations	Comments
CV	Rapid measurement	May not represent long-term activity	A combination of CV and chronoamperometry should be preferred over employing a single technique. CVs are relatively fast measurements, although slow scan rates should be favoured to avoid capacitive current contributions. Chrono-A measurements meanwhile could be kept relatively short to minimise the time of the experiment. Although some studies measure activities over rather large time periods, this can be considered the assessment of stability rather than intrinsic activity of the synthesised material.
	Easily extract Tafel slope region	Influenced by undefined capacitive currents	
Chrono-A	More indicative of steady-state activity	Multiple measurements required to extract Tafel slope region	
		Longer experiment time	
	No influence from capacitive currents	Possible (greater) influence from electrolyte impurities	
Chrono-P	Reveal potential oscillation processes	Requires pre-determined knowledge of I, V and time relationship. Difficult to perform in practice.	
Electrolyte			
H_2SO_4	Fewer issues with impurities	Adsorption of sulphates	Although HClO4 is sometimes employed in ORR studies, H_2SO_4 is largely employed in MOR studies. The only question is which concentrations should be favoured, often 0.5 mol dm^{-3} is used, above which the MOR is zero-order wrt. Methanol concentration.

(continued)

Table A.1 continued

Variable	Advantages	Limitations	Comments
$HClO_4$	Little adsorption of electrolyte (spectator) species	Measurement is particularly sensitive to impurities Electrolyte unstable at elevated temperature	
Temperature			
25 °C	Relevant to ambient DMFC devices	Measure: (i) Small currents (ii) Higher potentials	The choice of temperature should be based upon the intended DMFC conditions. For PtRu catalysts low Ru coverage is more active, whereas higher coverage are more active at 60 °C. Characterisation in both temperature ranges may therefore yield the most useful information.
60 °C	Measure: (i) Higher currents (ii) Lower potentials Relevant to elevated temperature DMFCs	Electrolyte stability Heated cell required Methanol evaporation	
Methanol concentration			
< 0.1 mol dm^{-3}	No obvious.	Partial MOR products may influence measurement.	Favour high MeOH concentrations. Reduced conversion of the MOR to CO_2 is less crucial because complete conversion is in general not achieved.
> 0.1 mol dm^{-3}	MOR on Pt is zero order wrt. MeOH conc. Closer to DMFC conditions.	Reduced conversion MOR to CO_2.	
Convection over electrode			
Rotation	Defined convection of MeOH to electrode surface	Lowered conversion of MeOH to CO_2	The MOR is largely partial oxidation and so conversion becomes of limited importance. Convection of MeOH to the electrode, however, is particularly important where low concentrations are employed.
No rotation	Higher conversion	Diffusion of MeOH to electrode can play a role	

(continued)

Table A.1 (continued)

Variable	Advantages	Limitations	Comments
Potential			
Fixed ca. 0.5 V_{RHE}	Fixed potential benchmark	Currents may be too small for certain materials to measure precisely (e.g. sensitive to current measurement offsets)	It will be difficult to compare many catalysts at a single potential. Instead, the comparison of catalysts at potentials where the Tafel-slope regions overlap would be appropriate. The Tafel-region at least indicates that the MOR activity limitations are kinetically limited, rather than by the blockage of the electrode.
Tafel region	Tafel slope identifies region where current scales with the RDS.	Requires multiple measurements, and there is no single point of comparison.	
Electrochemical pre-treatment			
Without pre-treatment	Residual impurities on the catalyst (from synthesis) may hinder the inherent activity of the catalyst.	Effects are not completely understood and changes the properties of the pristine synthesised material	Any pre-treatment on the electrode which influences the structure of the catalysts should be avoided. Although the activity may indeed be improved, this does not necessarily represent the intrinsic activity of the material synthesised. The irreversible structural effects of such pre-treatments are largely not understood and should rather be considered part of the catalyst synthesis and preparation procedure.
With pre-treatment	Electrocatalytic performance may be improved.		

Appendix B

Figures B.1, B.2, B.3, B.4.

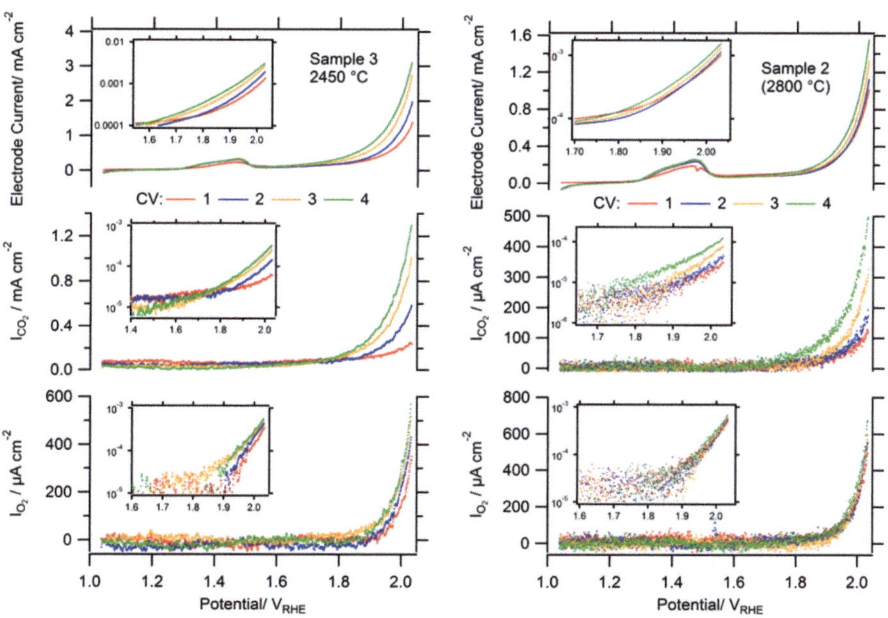

Fig. B.1 First four LSVs of the electrochemical COR of a heat treated carbon black Samples 3, 2 and 1 in 0.5 mol dm^{-3} H_2SO_4. Carbon black sample loadings of 85 ug$_c$ cm^{-2}on polycrystalline Au WE substrate

S. J. Ashton, *Design, Construction and Research Application of a Differential Electrochemical Mass Spectrometer (DEMS)*, Springer Theses,
DOI: 10.1007/978-3-642-30550-4, © Springer-Verlag Berlin Heidelberg 2012

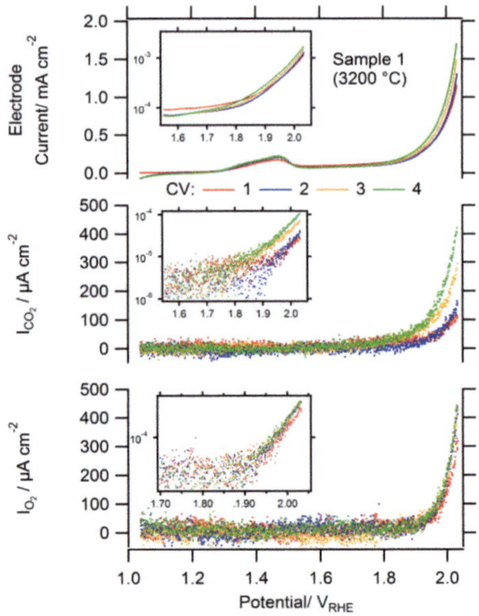

Fig. B.1 continued

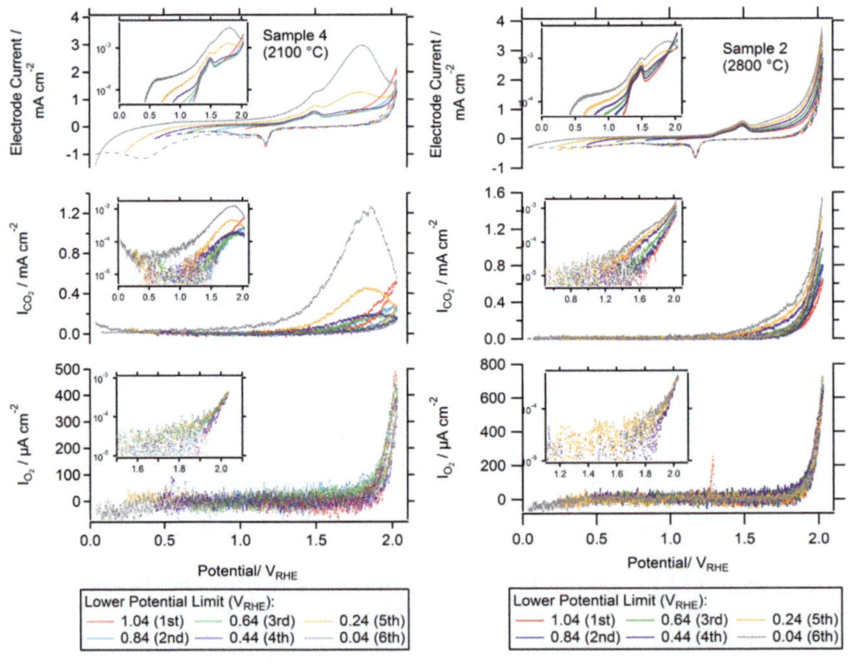

Fig. B.2 Effect of the lower potential limit on the electrochemical COR of the graphitised (and electrochemically oxidised in phase (ii) of the electrochemical treatment) Samples 2,4 and 5 in 0.5 mol dm^{-3} H_2SO_4. Sample loading of 85 ug$_c$ cm^{-2} on a polycrystalline Au WE substrate

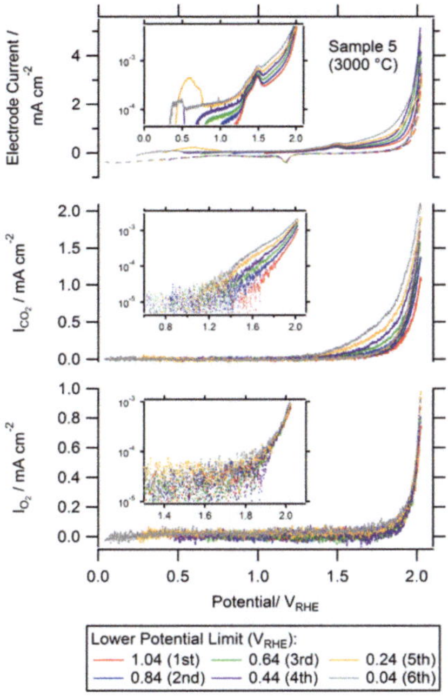

Fig. B.2 continued

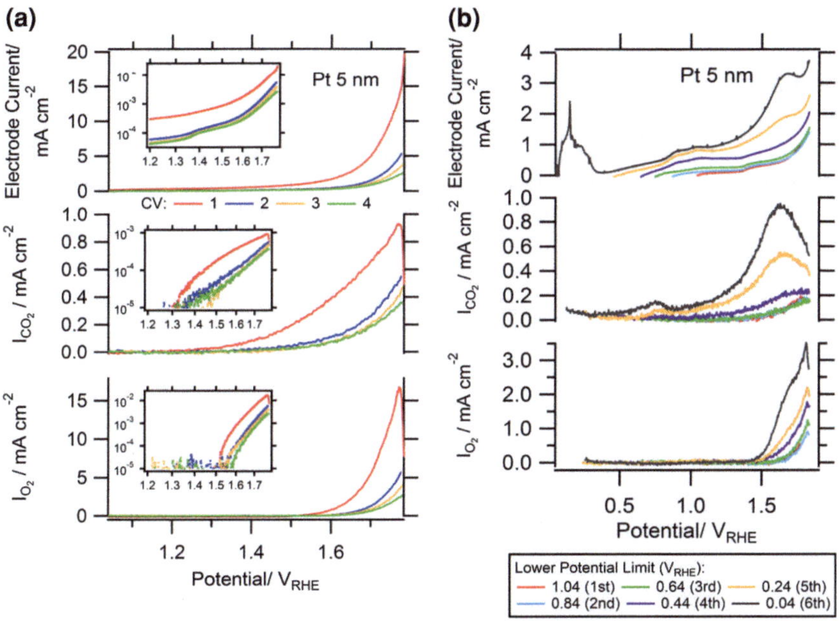

Fig. B.3 LSVs of repeated high potential excursion to 1.85 V_{RHE} on the electrochemical oxidation of the HSAC supported Pt 5 nm catalyst (**a**) and influence of the lower electrode potential limit following electrochemical oxidation (**b**)

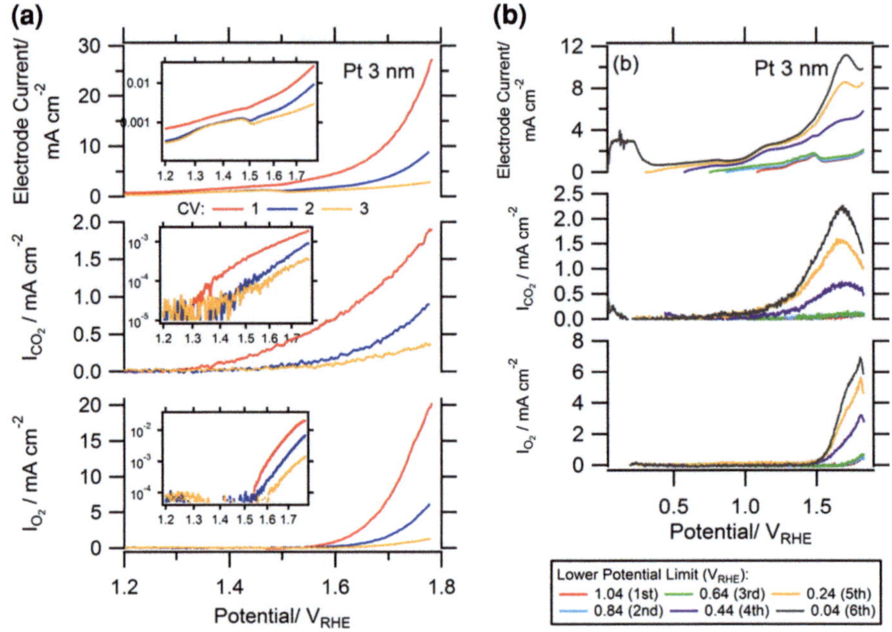

Fig. B.4 LSVs of repeated high potential excursion to 1.85 V_{RHE} on the electrochemical oxidation of the HSAC supported Pt 3 nm catalyst (**a**) and influence of the lower electrode potential limit following electrochemical oxidation (**b**)

References

1. Wiberg GK, Mayrhofer KJ, Arenz M (2009) Investigation of the oxygen reduction activity of non-platinum catalysts: a RDE methodology. ECS Meeting Abstracts 902(6): 349
2. Dillon R et al (2004) International activities in DMFC R&D: status of technologies and potential applications. J Power Sources 127(1–2): 112–126

Index

S. J. Ashton, *Design, Construction and Research Application of a Differential Electrochemical Mass Spectrometer (DEMS)*, Springer Theses,
DOI: 10.1007/978-3-642-30550-4, © Springer-Verlag Berlin Heidelberg 2012

MIX
Papier aus verantwortungsvollen Quellen
Paper from responsible sources
FSC® C105338

If you have any concerns about our products,
you can contact us on
ProductSafety@springernature.com

In case Publisher is established outside the EU,
the EU authorized representative is:
Springer Nature Customer Service Center GmbH
Europaplatz 3, 69115 Heidelberg, Germany

Printed by Libri Plureos GmbH
in Hamburg, Germany